I0429924

Common Bean Diseases

Daniel Diego Costa Carvalho

Common Bean Diseases, 2024.

How to cite this book:
CARVALHO, D.D.C. **Common Bean Diseases.** 1. Ed. Ipameri: Independently published. Kindle Direct Publishing, 2024. 87p.

Book description

This work contains current information on the etiology, symptomatology, epidemiology and control of the main bean diseases in Brazil:
Bean common mosaic
Bean yellow mosaic
Bean golden mosaic
Bacterial blight
Bacterial wilt
Anthracnose
Angular leaf spot
White mold
Powdery mildew
Fusarium wilt
Root-knot nematode
Cladosporium on seeds
Keywords: agriculture, agronomy, phytopathology, bean

Author's biography

Daniel Diego Costa Carvalho has degree in agronomy from the Federal University of Lavras (UFLA) and doctorate in phytopathology from the University of Brasília (UnB). He is currently a professor at the State University of Goiás (UEG).

Author's preface

The use of specialized literature composed of recent scientific articles and classic bibliographies outlined the construction of this work, wherein information is presented in a compact and dynamic form, allowing the reader to go beyond traditional literature. Below, the highlights for the fields of etiology, symptomatology, etiology and control, show a little of what can be appreciated in this work.

Etiology

This book proposes, for better understanding, the taxonomic positioning of the etiological agent within a complete and complex classification system. Concerning etiology, in the case of the Bacterial blight, a taxonomic review of species for the *X. axonopodis* complex stands out, suggesting the reclassification of these pathogens as *Xanthomonas phaseoli* pv. *phaseoli* and

Xanthomonas citri pv. *fuscans*

Symptomatology

Recent advances in the diagnosis of plant diseases make use of molecular techniques, most of the time not shared by the general public. In this sense, another point that deserves to be highlighted in view of the several limitations for the diagnosis of plant diseases lies in the better utilization of the diagnosis through precise and simplified symptoms characterization. As a practical example, this book mentions through text and image the most typical symptom found for bean anthracnose in the pods, where deep circular lesions are observed, among other commented characteristics.

Epidemiology

The epidemiology of plant diseases comprises in the study of a pathogen population regarding their action on a host population, that is, with a focus on the dynamics of this relationship, mainly involving space and time. Consequently, the epidemiology of this work has three main aspects: climatic conditions, dissemination and survival of the causal agent. An important point when referring to epidemiology is the fact that powdery mildew to be considered a useful model for studying the effects of climate change on plant diseases.

Control

Instead of simply citing or mentioning commonly used control measures, aiming in this work, in an innovative form and after an extensive review, what has been published in recent scientific research. White mold has several measures covered by six of the seven general control principles. In a didactic and compartmentalized form, the measures are commented and referenced one by one.

Acknowledgments

It is important to record my gratitude to the various support that I received from colleagues and professionals who make up my network, from the beginning of my student life to my recent professional life, with whom, helping me with valuable opinions and information and encouraging me with appreciation words and affection, make this aspiration a reality.

Online platform
Discover our platform for advanced studies on plant diseases:
https://sites.google.com/view/danieldiegocostacarvalho/

Comments
We do not make recommendations of any nature, the control examples described throughout this work are intended only for the scope of scientific and academic knowledge. All figures and images were designed and produced by the author.

Common Bean Diseases, 2024.

Summary

1. Bean common mosaic - *Bean common mosaic virus*   1

2. Bean yellow mosaic - *Bean yellow mosaic virus*   7

3. Bean golden mosaic - *Bean golden mosaic virus*   13

4. Bacterial blight - *Xanthomonas phaseoli* **pv.** *phaseoli,*
Xanthomonas citri **pv.** *fuscans, Xanthomonas cannabis* **pv.** *phaseoli*
    19

5. Bacterial wilt - *Curtobacterium flaccumfaciens* **pv.**
flaccumfaciens   27

6. Anthracnose - *Colletotrichum lindemuthianum*   35

7. Angular leaf spot - *Pseudocercospora griseola*   41

8. White mold - *Sclerotinia sclerotiorum*   49

9. Powdery mildew - *Erysiphe polygoni*   59

10. Fusarium wilt - *Fusarium oxysporum* **f. sp.** *phaseoli*   65

11. Root-knot nematode - *Meloidogyne* **spp.**   75

12. Cladosporium on seeds - *Cladosporium herbarum*   83

Common Bean Diseases, 2024.

1. Bean common mosaic - *Bean common mosaic virus*

Family: Potyviridae
Genus: Potyvirus

Etiology

The nomenclature and classification of viruses is a prerogative of the International Committee on Taxonomy of Viruses (ICTV). The organization of taxa within a classification system is based on properties, including the main ones: particle morphology, type of nucleic acid, genome organization, replication and serology. Regarding the name of a viral species, the ICTV established that the common name of the virus in the English language is considered as the scientific name of the viral species. Normally, the name of the virus is composed of the name of the plant where the virus was found plus the type of symptoms that it causes.

The species *Bean common mosaic virus* comprises in a positive-sense single-stranded RNA [(+)ssRNA] genome, that is, this virus acts directly as a messenger RNA (mRNA) (ZAMORA et al., 2017). Once in the plant cell cytoplasm, the virus loses its protein coat (decapsidation) and releases the genetic material, which becomes available for events that aimed at producing new viral particles. In short, the virus will use the plant cell's metabolic machinery to produce the products of its interest, such as coat proteins.

The encapsidation of the potyviral RNA genome occurs within the coat protein core, which forms a characteristic flexuous rod-shaped structure measuring 11-20 nm in diameter with 680-900 nm length (ZAMORA et al., 2017; KUMAR et al., 2019). Another important sign of the *Bean common mosaic virus* comprises in the formation of pinwheel-like cylindrical inclusions within the cytoplasm of infected plant cells (AGRIOS, 2005).

Bean common mosaic virus belongs to the Family Potyviridae,

Genus Potyvirus (Figure 1). According to JORDAN & HAMOND (2008), this virus infects approximately 100 species of 44 genera of nine plant families.

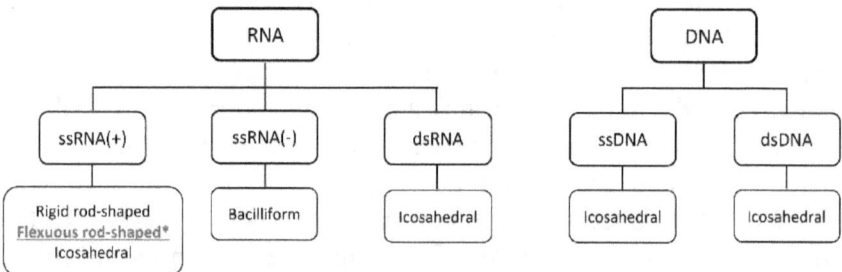

Figure 1. Schematic diagram shows the positioning of *Bean common mosaic virus* within virus systematic based on the genetic material and the morphology of the viral particle.

Symptomatology

The main characteristic of the bean common mosaic regarding the symptoms is the appearance of light-green areas alternating with dark-green, leaflets twisting, blistering and leaflets shrinkage (Figure 2). The occurrence of necrosis is not common and is conditioned to the specific occasions such as the prevalence of serotype A strains, varieties that have the hypersensitivity gene and specific environmental conditions (JORDAN & HAMMOND, 2008).

Figure 2. Bean common mosaic: leaflets twisting, blistering and leaflets shrinkage.

Epidemiology

Climate conditions: Favorable climatic conditions for the vector insects must be considered at the planning moment, since these conditions are, consequently, potential for the disease occurrence. The temperature of 25°C provided the best thermal condition for the population growth of *Myzus persicae* in pepper (BARBOSA et al., 2011).

Dissemination: The transmission of *Bean common mosaic virus* occurs by mechanical contact, seeds and insects. Regarding to transmission by insects, the virus-vector interrelationship is of the non-circulating non-persistent type, that is, the virus does not circulate inside the insect vector. This virus is restricted to the mouthparts of the insect vector, whereas the viral particle remains and, therefore, viable until the first feeding of the insect vector (WENDLAND et al., 2016). The main species of insect vectors are: *Acyrthosiphon pisum, Aphis fabae, Myzus persicae* and *Aphis* *craccivora.*

Survival: The *Bean common mosaic virus* has many plant species to act as alternative hosts, such as species of the genus *Phaseolus* (*P. acutifolius* and *P. coccineus*) *Macroptilium atropurpureus, Glycine max, Pisum sativum, Rhynchosia minima, Vigna mungo, V. angularis* and *V. unguiculata* (JORDAN & HAMOND, 2008).

Control
Evasion: When it comes to control, measures should focus on reducing the primary inoculum. Seeds must be sown in growing areas far from alternative hosts.

Exclusion: As the virus is transmitted by seeds, another control measure comprises in the use of virus-free seeds. The preventive application of insecticides is unfeasible (JORDAN & HAMOND, 2008).

Immunization: Normally, the virus will be present in seed material that no have the dominant I gene for resistance to *Bean common mosaic virus*. In this regard, another fundamental measure consisting of the use of resistant varieties.

References
AGRIOS, G.N. **Plant Pathology.** 5th ed. San Diego: Academic Press, 2005, 922p.

BARBOSA, L.R.; CARVALHO, C.F.; AUAD, A.M.; SOUZA, B.; BATISTA, E.S.P. Tabelas de esperança de vida e fertilidade de Myzus persicae sobre pimentão em laboratório e casa de vegetação. **Bragantia,** v.70, n.2, p.375-382, 2011. https://doi.org/10.1590/S0006-87052011000200018

JORDAN, R.; HAMMOND, J. *Bean common mosaic virus* **and** *Bean common mosaic necrosis virus*. Encyclopedia of Virology, p.288-295, 2008.

KUMAR, S.; KARMAKAR, R.; GARG, D.K.; GUPTA, I.; PATEL, A.K. Elucidating the functional aspects of different domains of *Bean common mosaic virus* coat protein. **Virus Research**, v.273, e197755, 2019. https://doi.org/10.1016/j.virusres.2019.197755

WENDLAND, A.; MOREIRA, A.S.; BIANCHINI, A.; GIAMPAN, J.S.; LOBO JUNIOR, M. Doenças do Feijoeiro. In: AMORIM, L.; REZENDE. J.A.M.; BERGAMIN FILHO, A.; CAMARGO, L.E.A. **Manual de Fitopatologia: Doenças das plantas cultivadas**. vol.2, 5.Ed. Ouro Fino: Agronômica Ceres, pp.383-396, 2016.

ZAMORA, M.; MÉNDEZ-LÓPEZ, E.; AGIRREZABALA, X.; CUESTA, R.; LAVÍN, J.L.; SÁNCHEZ-PINA, M.A.; ARANDA, M.A.; VALLE, M. Potyvirus virion structure shows conserved protein fold and RNA binding site in ssRNA viruses. **Science Advances**, v.3, n.9, eaao2182, 2017. https://doi.org/10.1126/sciadv.aao2182

2. Bean yellow mosaic - *Bean yellow mosaic virus*

Family: Potyviridae
Genus: Potyvirus

Etiology
As in the case of *Bean common mosaic virus*, the species *Bean yellow mosaic virus* is also a flexuous rod-shaped virus (SCHULZE et al., 2017) and has a genetic material of the positive-sense single-stranded RNA type [(+)ssRNA].

The *Bean yellow mosaic virus* is serologically related to the *Bean common mosaic virus* (DRIJFHOUT & BOS, 1977). As for ultrastructural characters, RADWAN et al. (2008) verified in transmission electron microscopy filamentous particles of *Bean yellow mosaic virus* measuring about 750 µm long and 14 µm wide. Furthermore, inclusions can be found in the cytoplasm of infected plant cells. In *Vicia faba*, inclusions in straight opaque stripes or slightly curved and opaque crystals of varying shapes, often hexagonal have been reported by RADWAN et al. (2008). There are several deposits of *Bean yellow mosaic virus* sequences on GenBank platform, which facilitate the verification of nucleotide identity. An example is accession KX907126, found in the alternative host *Lupinus albus* (SCHULZE ct al., 2017).

In a complete classification, *Bean yellow mosaic virus* belongs to the phylum Pisuviricota, class Stelpaviricetes, order Patatavirales, family Potyviridae, genus Potyvirus. For this book, regarding this topic and, when it comes to viruses, attention will be directed to the type of genetic material, the morphology of viral particle (Figure 3), the family and the genus.

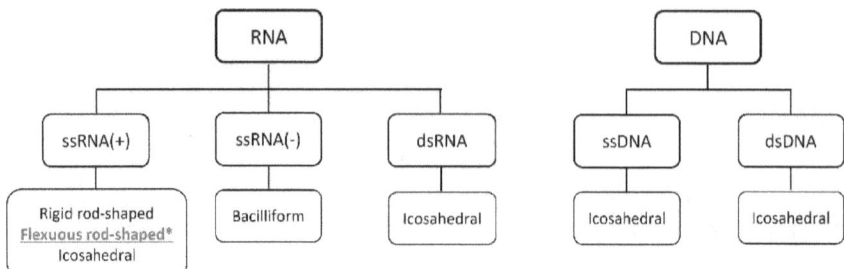

Figure 3. Schematic diagram shows the positioning of *Bean yellow mosaic virus* within virus systematic based on the genetic material and the morphology of the viral particle.

Symptomatology

The most common symptom of the bean yellow mosaic is a yellow-green mosaic on the leaves, more severe than bean common mosaic (Figure 4), in which blistering, foliar wrinkles and leaflets shrinkage may occur (WENDLAND et al., 2016).

Figure 4. Bean yellow mosaic: yellow-green mosaic on the leaves, more severe and more yellow than the bean common mosaic.

Epidemiology

Climate conditions: Air temperature and rainfall are the climatic elements with the greatest direct and indirect impacts on insect populations (ROSADO et al., 2015; SOARES et al., 2020). In general, the average development time of these organisms decreases with increasing temperature within the thermal range required for the survival of each species (CAMPBELL & MACKAUER, 1975). Favorable climatic conditions for vector insects increase the occurrence of the disease and must be observed.

Dissemination: Although there are reports that the *Bean yellow mosaic virus* is not transmitted by common bean seeds, the fact that it is transmitted by other Fabaceae seeds must be considered (SOFY et al., 2020), since these plants can occur in crops as alternative hosts. Besides, *Bean yellow mosaic virus* is transmitted non-persistently mode by aphids (RADWAN et al., 2008).

Survival: Among the various host plant for this virus, it is included other Fabaceae species such as *Kennedia prostrata, Vicia faba, Lupinus angustifolius, L. luteus, L. pilosus, L. albus, Melilotus indica* and *Pisum sativum* (WYLIE et al., 2008).

Control

The bean yellow mosaic is difficult to control due to its wide spectrum of alternative hosts and the non-persistent interrelation with the insect vector. Besides, the number of chemical pesticides used in the management of plant viral diseases is limited (SOFY et al., 2020). There are no effective methods for controlling bean yellow mosaic.

Immunization: In the bibliography there are experimental results regarding the use of salicylic acid (RADWAN et al., 2008) and carboxymethyl chitosan-titania nanobiocomposites aiming to trigger the *Vicia faba* defense system against to *Bean yellow mosaic virus* (SOFY et al., 2020).

References

CAMPBELL, A.; MACKAUER, M. Thermal constants for development of the pea aphid (Homoptera: Aphididae) and same of its parasites. **The Canadian Entomologist**, v.107, n.4, p.419-423, 1975.

DRIJFHOUT, E.; BOS, L. The identification of two new strains of bean common mosaic vírus. **Netherlands Journal of Plant Pathology**, v.83, p.13-25, 1977. https://doi.org/10.1007/BF01976508

RADWAN, D.E.M.; LU, G.; FAYEZ, K.A.; MAHMOUD, S.Y. Protective action of salicylic acid against *Bean yellow mosaic virus* infection in *Vicia faba* leaves. **Journal of Plant Physiology**, v.165, p.845-857, 2008. https://doi.org/10.1016/j.jplph.2007.07.012

ROSADO, J.F.; PICANÇO, M.C.; SARMENTO, R.A.; SILVA, R.S.; PEDRO-NETO, M.; CARVALHO, M.A.; ERASMO, E.A.L.; SILVA, L.C.R. Seasonal variation in the populations of *Polyphagotarsonemus latus* and *Tetranychus bastosi* in physic nut (*Jatropha curcas*) plantations. **Experimental and Applied Acarology,** v.66, p.415-426, 2015. https://doi.org/10.1007/s10493-015-9911-6

SCHULZE, A.; ROBERTS, R.; PIETERSEN, G. First Report of the Detection of *Bean yellow mosaic virus* (BYMV) on *Tropaeolum majus*; *Hippeastrum* spp., and *Liatris* spp. in South Africa. **Plant Disease**, v.101, n.5, p.846, 2017. https://doi.org/10.1094/PDIS-10-16-1446-PDN

SOARES, J.R.S.; PAES, J.S.; ARAÚJO, V.C.R.; ARAÚJO, T.A.; RAMOS, R.S.; PICANÇO, M.C.; ZANUNCIO, J.C. Spatiotemporal Dynamics and Natural Mortality Factors of *Myzus persicae* (Sulzer) (Hemiptera: Aphididae) in Bell Pepper Crops. **Neotropical Entomology**, v.49, p.445-455, 2020. https://doi.org/10.1007/s13744-020-00761-2

SOFY, A.R.; HMED, A.A.; ALNAGGAR, A.E.M.; DAWOUD, R.A.; ELSHAARAWY, R.F.M.; SOFY, M.R. Mitigating effects of *Bean yellow mosaic virus* infection in faba bean using new carboxymethyl chitosan-titania nanobiocomposites. **International Journal of Biological Macromolecules,** v.163, p.1261-1275, 2020. https://doi.org/10.1016/j.ijbiomac.2020.07.066

WENDLAND, A.; MOREIRA, A.S.; BIANCHINI, A.; GIAMPAN, J.S.; LOBO JUNIOR, M. Doenças do Feijoeiro. In: AMORIM, L.; REZENDE. J.A.M.; BERGAMIN FILHO, A.; CAMARGO, L.E.A. **Manual de Fitopatologia: Doenças das plantas cultivadas.** vol.2, 5.Ed. Ouro Fino: Agronômica Ceres, pp.383-396, 2016.

WYLIE, S.J.; COUTTS, B.A.; JONES, M.G.K.; JONES, R.A.C. Phylogenetic analysis of *Bean yellow mosaic virus* isolates from four continents: Relationship between the seven groups found and their hosts and origins. **Plant Disease,** v.92, n.12, p.1596-1603, 2008. https://doi.org/10.1094/PDIS-92-12-1596

3. Bean golden mosaic - *Bean golden mosaic virus*

Family: Geminiviridae
Genus: Begomovirus

Etiology

According to ZERBINI et al. (2017), the family Geminiviridae has the characteristics an icosahedral particle composed of two almost isometric germinated subunits, each containing a single-stranded circular DNA (DNA-A and DNA-B) as genetic material, respectively.

Each germinated isometric particles measure 22 to 38 nm in diameter. Such particles can only be seen in transmission electron microscopy. According to SNEHI et al. (2017), the genome of the genus Begomovirus is organized into ORFs (Open Reading frame), which are intervals of DNA sequence, which encode a product with a predicted function within the viral cycle. In the specific case, according to the mentioned authors, the ORFs AV1, AV2, AC1, AC2, AC3 and AC4 are located in the DNA-A, which code for products related to the encapsidation, cell-to-cell movement protein, replication initiation, transcription activators of rightward ORFs, replication enhancement and PTGS (post-transcriptional gene silencing) suppressor, respectively. The BV1 and BC1 ORFs are located in the DNA-B, which encode products related to nuclear trafficking (NSP) and cell-to-cell movement – determinants of pathogenicity, respectively. Thus, after presenting the predicted functions of the different ORFs, it is noted that for the virus to be infective, it requires both DNA-A and DNA-B components.

The taxonomic positioning of the *Bean golden mosaic virus* in a simplified perspective can be seen in figure 5.

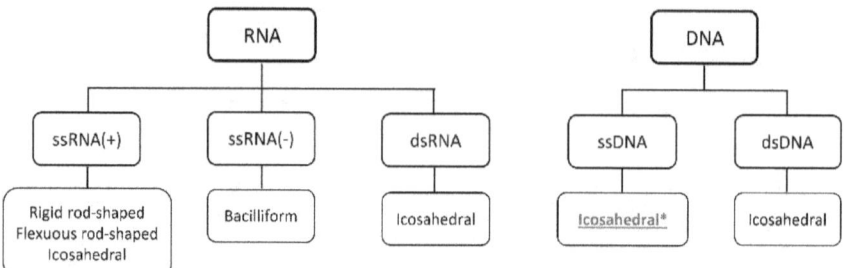

Figure 5. Schematic diagram shows the positioning of *Bean golden mosaic virus* within virus systematic based on the genetic material and the morphology of the viral particle.

Symptomatology

The main symptom is characterized by a bright yellow-green mosaic on the leaves, stunted growth and distorted pods (ARAGÃO et al., 2013; BATISTA et al., 2022). Symptoms begin in the first trifoliate leaves, always associated with the presence of the vector *Bemisia tabaci*, in which intense leaf yellowing occurs, followed by other important markers such as leaf curling, wrinkling, veins chlorosis, dwarfism, oversprouting and senescence retardation (WENDLAND et al., 2018). To assist in diagnosing the disease, it is important to use PCR with species-specific primers, as described by BONFIM et al. (2007).

Epidemiology

Climate conditions: *Bemisia tabaci* is the insect vector of the bean golden mosaic virus, whose virus-vector interrelationship is of the persistent circulative type (SNEHI et al., 2017). Concerning to *Bemisia tabaci*, in economically important crops, studies indicate that the development time of this insect decreases as there is an increase in the temperature from 14.9°C to 30.0°C in cotton crops (BUTLER et al., 1983). In another study, conducted on tomato crops, BONATO et al. (2007), reported that the development rate of *B. tabaci* increases, especially between 17°C and 30°C. These favorable climatic conditions must be observed when carrying out the culture in question, since the functional relationships between temperature and insect life are used to evaluate the effect on population dynamics (BONATO et al., 2007).

Dissemination: Bean golden mosaic virus, unlike common bean mosaic virus, is not transmitted by the seeds or via mechanical transmission.

Survival: According to BATISTA et al. (2022), the genus Begomovirus has been reported in several hosts such as *Phaseolus vulgaris*, *P. lunatus*, *Glycine max*, *Vigna unguiculata*, *Calopogonium* sp., *Desmodium* sp., and *Macroptilium* sp.

Control
Eradication: Most control measures aim to reduce the initial inoculum. Thus, the sanitary void adopted for beans is mentioned as an effective measure (WENDLAND et al., 2018). Briefly, control measures are mainly focused for controlling the insect vector, more specifically on monitoring its population. However, problems arising from the use of insecticides have made the task difficult, among which we can mention the development of the insecticide insect resistance, low cost-benefit ratio and concerns about the environment (BONFIM et al., 2007). Attention must be paid to the alternative hosts mentioned in the previous section, in the sense of to carry out distant planting or even the elimination of these insect vector hosts.

Immunization: The corporation Empresa Brasileira de Pesquisa Agropecuária (Embrapa) developed the Embrapa 5.1 bean plant through genetic transformation. This is a material whose defense mechanism is post-transcriptional gene silencing against specific sequences of mRNA transcribed from the AC1 (rep) gene of the *Bean golden mosaic virus* DNA-A. The rep gene is a gene involved in necessary and sufficient functions for viral replication. In this mechanism, the rep gene is inactivated via silencing of the transcribed mRNA, which makes it impossible for the virus to replicate. As a result, the plant becomes resistant (FARIA & ARAGÃO, 2013).

References

ARAGÃO, F.J.L.; NOGUEIRA, E.O.P.L.; TINOCO, M.L.P.; FARIA, J.C. Molecular characterization of the first commercial transgenic common bean immune to the *Bean golden mosaic virus.* **Journal of Biotechnology,** v.166, p.42-50, 2013. http://dx.doi.org/10.1016/j.jbiotec.2013.04.009

BATISTA, J.G.; NERY, F.M.B.; MELO, F.F.S.; MALHEIROS, M.F.; REZENDE, D.V.; BOITEUX, L.S.; FONSECA, M.E.N.; MIRANDA, B.E.C.; PEREIRA-CARVALHO, R.C. Complete genome sequence of a novel bipartite begomovirus infecting the legume weed *Macroptilium erythroloma.* **Archives of Virology,** v.167, p.1597-1602, 2022. https://doi.org/10.1007/s00705-022-05410-0

BONATO, O.; LURETTE, A.; VIDAL, C.; FARGUES, J. Modelling temperature-dependent bionomics of *Bemisia tabaci* (Q-biotype). **Physiological Entomology,** v.32, p.50-55, 2007. https://doi.org/10.1111/j.1365-3032.2006.00540.x

BONFIM, K.; FARIA, J.C.; NOGUEIRA, E.O.P.L.; MENDES, E.A.; ARAGÃO, F.J.L. RNAi-mediated resistance to *Bean golden mosaic virus* in genetically engineered common bean (*Phaseolus vulgaris*). **Molecular Plant-Microbe Interactions,** v.20, n.6, p.717-726, 2007. https://doi.org/10.1094/MPMI-20-6-0717

BUTLER, G.D.; HENNEBERRY, T.J.; CLAYTON, T.E. *Bemisia tabaci* (Homoptera: Aleyrodidae): Development, Oviposition, and Longevity in Relation to Temperature. *Annals of the Entomological Society of America*, v.76, n.2, p.310-313, 1983. https://doi.org/10.1093/aesa/76.2.310

FARIA, J.C.; ARAGÃO, F.J.L. **Embrapa 5.1: o feijoeiro geneticamente modificado resistente ao mosaico dourado.** Empresa Brasileira de Pesquisa Agropecuária, Embrapa Arroz e Feijão, Ministério da Agricultura, Pecuária e Abastecimento. Santo Antônio de Goiás: Embrapa Arroz e Feijão. Documentos 291, 2013. 48p.

SNEHI, S.K.; PURVIA, A.S.; PARIHAR, S.S.; GUPTA, G.; SINGH, V.; RAJ, S.K. Overview of Begomovirus genomic organization and its impact. **International Journal of Current Research,** v.9, n.1, p.61368-61380, 2017.

WENDLAND, A.; LOBO JUNIOR, M.; FARIA, J.C. **Manual de identificação das principais doenças do feijoeiro-comum.** Empresa Brasileira de Pesquisa Agropecuária, Embrapa Arroz e Feijão, Ministério da Agricultura, Pecuária e Abastecimento. Brasília: Embrapa, 2018. 49p.

ZERBINI, F.M.; BRIDDON, R.B.; IDRIS, A.; MARTIN, D.P.; MORIONES, E.; NAVAS-CASTILLO, J.; RIVERA-BUSTAMANTE, R.; ROUMAGNAC, P.; VARSANI, A. ICTV Virus Taxonomy Profile: Geminiviridae. **Journal of General Virology,** v.98, p.131-133, 2017. https://doi.org/10.1099/jgv.0.000738

4. Bacterial blight - *Xanthomonas phaseoli* pv. *phaseoli*, *Xanthomonas citri* pv. *fuscans*, *Xanthomonas cannabis* pv. *phaseoli*

Domain: Bacteria
Phylum: Proteobacteria
Class: Gammaproteobacteria
Order: Xanthomonadales

Etiology
The systematics of bacteria is in constantly changing. New descriptions follow the rules of the International Code of Nomenclature of Prokaryotes and, for pathovars, follow the rules of the International Standards for Naming Pathovars. Any name change or new description must be published exclusively in the International Journal of Systematic and Evolutionary Microbiology (BEDENDO & BELASQUE, 2018).

The genus *Xanthomonas* has the characteristics to be a gram-negative bacteria, that is, it has a thinner peptidoglycan layer compared to gram-positive bacteria, it stains red in the Gram test and presents the production of mucus when on the KOH test.

The bacteria of the genus *Xanthomonas* measure 0.4-1,0 x 1.2-3.0 µm (AGRIOS, 2005), have a bacilliform defined format and polar flagellum (monotrichous). The species of *Xanthomonas*, causal agents of the common bacterial blight of beans, have some colony morphology, physiological and biochemical characteristics, which can assist in identification (MAHUKU et al., 2006). However, genetic characterization using the multiplex PCR technique made it possible to separate not only the species *Xanthomonas phaseoli* pv. *phaseoli* and *Xanthomonas citri* pv. *fuscans*, but also the identification of currently known lineages (PAIVA et al., 2022).

The taxonomy of infective strains has been debated since the identification of the genus *Xanthomonas* as causal agent of the bacterial blight in 1987. It is not uncommon to find bibliographs referring to the common bean bacterial blight causal agents as *Xanthomonas axonopodis* pv. *phaseoli* and *Xanthomonas fuscans* subsp. fuscans. CONSTANTIN et al. (2016) proposed a taxonomic revision of species for the *X. axonopodis* complex, suggesting the reclassification of these pathogens as *Xanthomonas phaseoli* pv. *phaseoli* and *Xanthomonas citri* pv. *fuscans*. From a collection of 117 strains of *Xanthomonas* isolated from common bean plants from various producing regions in Brazil, phylogenetic analyzes carried out by PAIVA et al. (2022) revealed that all genetic variants of the pathogens responsible for common bean bacterial blight (NF1, NF2, NF3 and fuscans) are present in Brazil, presenting significant virulence variability. Specifically, these detected genetic lineages are distributed between the two species: *Xanthomonas citri* pv. *fuscans* (fuscans lineage, NF2 and NF3) and *Xanthomonas phaseoli* pv. *phaseoli* (NF1 lineage). In addition to these mentioned species, the bacteria *Xanthomonas cannabis* pv. *phaseoli* is also considered the causal agent of common bean bacterial blight (PAIVA, 2018).

The genus *Xanthomonas* is positioned in the Gammaproteobacteria class (Figure 6), the same class of other important phytobacteria, such as *Pseudomonas*, *Erwinia* and *Pectobacterium*.

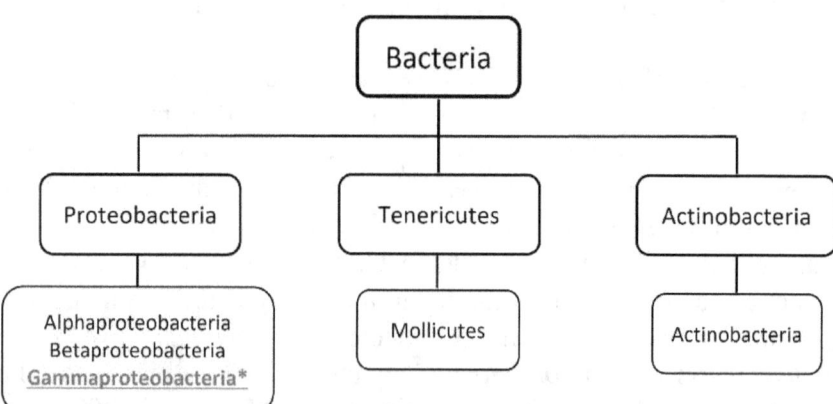

Figure 6. Schematic diagram shows the positioning of the genus *Xanthomonas* within the recent systematics of bacteria. The Phylum Actiniobacteria groups Gram-positive bacteria, while the Phylum Proteobacteria contains Gram-negative bacteria. The Phylum

Tenericutes includes bacteria lacking cell wall.

Symptomatology

The species *Xanthomonas phaseoli* pv. *phaseoli* and *Xanthomonas citri* pv. *fuscans* involved with this pathosystem cause similar symptoms, attacking leaves, stems, pods and seeds (SILVA JÚNIOR et al., 2022). During the pathogenesis process, the bacteria colonize the intercellular spaces, where there is intercellular fluid, a liquid medium rich in nutrients, constituting an excellent culture medium. In this fluid, the bacteria begin to multiply and moves between cells (ROMEIRO, 2005). Due to this, the lesions acquire characteristics that initially refer to small lesions with a soaked appearance of the tissue. In a more advanced condition, lesions increase and coalesce. The coalesced lesions begin to form necrotic tissue with a yellowish halo (Figure 7). As the process continues, these lesions can occupy the entire length of the leaf blade and with even larger yellow halos.

It is worth mentioning here the unusual symptoms observed by the strain of *Xanthomonas* called "Nyagatare", found in Rwanda, on the African continent, which are characterized by bean plants exhibiting leaf curling and wilting, brownish and white spots and brown to dark necrosis on the veins (ARITUA et al. 2015). In this same study, after genome sequencing, the Nyagatare strain was positioned in a clade different from that of *Xathomonas axonopodis*, X. *citri* and X. *fuscans*, establishing that the "Nyagatare" strain was not phylogenetically related to the other associated species to common bacterial blight.

Figure 7. Common bean bacterial blight: initial lesions showing necrotic tissue with a yellowish halo.

Epidemiology
Climate conditions: The occurrence of common bean bacterial blight depends of high temperatures and high humidity (SILVA JÚNIOR et al., 2022).

Dissemination: Bacterial attack is based on passive penetration through natural openings (stomata, lenticels and hydathodes) and wounds. When it comes to penetration via stomata, it is important to highlight that there is a need for a leaf water film and high relative humidity in the sub-stomatal chamber for bacterial infections to proceed. Spread over short distances occurs with the help of rainwater, while seeds are responsible for spreading over long distances.

Survival: After inciting symptoms, the bacteria survive for long periods in cultural remains, seeds and alternative hosts. According to SILVA JÚNIOR et al. (2022), at least 41 plant species, from economically important crops to weeds, have been reported as alternative hosts for *Xanthomonas phaseoli* pv. *phaseoli* and *Xanthomonas citri* pv. *fuscans*, whether by natural infection or artificial inoculation: *Acalypha alopecuroidea, Acanthospermum hispidum, Aeschynomene americana, Amaranthus retroflexus, Ambrosia artemisiifolia, Beta vulgaris, Calopogonium* sp., *Cenchrus echinatus, Chenopodium album, Cyperus rotundus, Digitaria sclalarum, Echinochloa colona, E. crus-galii, Euphorbia heterophylla, Glycine max, Lablab purpureus, Leptochloa filiformis, Lupinus polyphyllus, Macroptilium lathyroides, Malachra alceifolia, Mucuna deeringiana, Phaseolus acutifolius, P. coccineus, P. lunatus, Physalis* sp., *Pisum sativum, Portulaca oleracea, Pueraria* sp., *Rhynchosia minima, Ruellia tuberosa, Senna hirsuta, Solanum nigrum, Strophostyles helvola, Vicia sativa, V. villosa, Vigna aconitifolia, V. angularis, V. mungo, V. radiata, V. umbellata* and *V. unguiculata*.

Control
Regarding cultural practices, the measures will focus on creating unfavorable conditions for the pathogen and interfering with its survival, reproduction and dissemination.

Evasion: It is recommended to plant crops 30 meters away from other crops.

Exclusion: To carry out the sowing of bacteria-free seeds.

Eradication: Among the measures within the eradication principle, the following are recommended: crop rotation, removal of alternative hosts and elimination of crop remains.

Regulation: Employ adequate fertilization and irrigation and avoid traffic when the plants are wet (BEDENDO et al., 2018). In Brazil, there are no commercial products registered for the biological control of common bean bacterial blight. However, several genus of biocontrol agents, mainly bacteria, have the potential for the development of commercial products aimed at controlling this

23

disease (SILVA JÚNIOR et al., 2022).

Immunization: When it comes to common bean bacterial blight, the main control measure is based on the use of resistant cultivars. In Brazil, the cultivars BRS Notável (carioca beans) and BRS Esplendor (black beans) are considered as resistant. The other cultivars available on the market are considered moderately resistant, including cultivars from the special group (Jalo and Rajado beans) (SILVA JÚNIOR et al., 2022).

Therapy: Regarding the use of chemical products, there are studies concerning to the application of antibiotics on seed treatment. However, there are unsatisfactory results and without complete eradication of the bacteria (LIANG et al., 1992). In crops, application of copper sulfate or copper hydroxide is indicated for the management of common bean bacterial blight (BELETE & BATAS, 2017).

References

AGRIOS, G.N. **Plant Pathology.** 5th ed. San Diego: Academic Press, 2005, 922p.

ARITUA, V.; MUSONI, A.; KABEJA, A.; BUTARE, L.; MUKAMUHIRWA, F.; GAHAKWA, D.; KATO, F.; ABANG, M.M.; BURUCHARA, R.; SAPP, M.; HARRISON, J.; STUDHOLME, D.J.; SMITH. J. The draft genome sequence of *Xanthomonas* species strain Nyagatare, isolated from diseased bean in Rwanda. **FEMS Microbiology Letters**, v.362, n.4, fnu055, 2015. https://doi.org/10.1093/femsle/fnu055

BEDENDO, I.P.; BELASQUE, J. Bactérias fitopatogênicas. In: AMORIM, L.; REZENDE, J.A.M.; BERGAMIN FILHO, A. **Manual de Fitopatologia: princípios e conceitos**. vol.1, 5.Ed. Ouro Fino: Agronômica Ceres, pp.143-160, 2018.

BELETE, T.; BASTAS, K.K. Common bacterial blight (*Xanthomonas axonopodis* pv. *phaseoli*) of beans with special focus on Ethiopian condition. **Journal of Plant Pathology & Microbiology**, v.8, p.1-10, 2017. https://doi.org/10.4172/2157-7471.1000403

CONSTANTIN, E.C.; CLEENWERCK, I.; MAES, M.; BAEYEN, S.; VAN MALDERGHEM, C.; DE VOS, P.; COTTYN, B. Genetic characterization of strains named as *Xanthomonas axonopodis* pv. *dieffenbachiae* leads to a taxonomic revision of the *X. axonopodis* species complex. **Plant Pathology**, v.65, p.792-806, 2016. https://doi.org/10.1111/ppa.12461

LIANG, L.Z.; HALLOIN, J.M.; SAETTLER, A.W. Use of polyethylene glycol and glycerol as carriers of antibiotics for reduction of *Xanthomonas campestris* pv. *phaseoli* in navy bean seeds. **Plant Disease**, v.76, p.875-879, 1992. https://doi.org/10.1094/PD-76-0875

MAHUKU, G.S.; JARA, C.; HENRIQUEZ, M.A.; CASTELLANOS, G.; CUASQUER, J. Genotypic Characterization of the common bean bacterial blight pathogens, *Xanthomonas axonopodis* pv. *phaseoli* and *Xanthomonas axonopodis* pv. *phaseoli* var. *fuscans* by rep-PCR and PCR–RFLP of the ribosomal genes. **Journal of Phytopathology**, v.154, p.35-44, 2006. https://doi.org/10.1111/j.1439-0434.2005.01057.x

PAIVA, B.A.R. **Crestamento bacteriano do feijoeiro no Brasil: distribuição, diversidade e detecção de seus agentes causais *Xanthomonas* spp.** Tese (Doutorado em Fitopatologia). Departamento de Fitopatologia, Universidade de Brasília, Brasília, 2018. 176p.

PAIVA, B.A.R.; WENDLAND, A.; ROSSATO, M.; FERREIRA, M.A.S.V. Virulence and type III effector diversities of *Xanthomonas citri* pv. *fuscans* and X. *phaseoli* pv. *phaseoli* in Brazil. **Journal of Phytopathology**, v.170, p.1-14, 2022. https://doi.org/10.1111/jph.13049

ROMEIRO, R.S. **Bactérias fitopatogênicas,** 2 ed. Viçosa: Editora UFV. 2005. 417p.

SILVA JÚNIOR, T.A.F.; NASCIMENTO, D.M.; SILVA, J.C.; SOMAN, J.M.; GONÇALVES, R.M.; MARINGONI, A.C. Common bacterial blight of beans: an integrated approach to disease management in Brazil. **Tropical Plant Pathology**, v.47, p.457-469, 2022. https://doi.org/10.1007/s40858-022-00504-1

5. Bacterial wilt - *Curtobacterium flaccumfaciens* pv. *flaccumfaciens*

Domain: Bacteria
Phylum: Actinobacteria
Class: Actinobacteria
Order: Micrococcales

Etiology

Among some main morphological characteristics, it is notable in Gram-positive bacteria that they have a thick layer of peptidoglycan when compared to Gram-negative bacteria. The rigid peptidoglycan layer gives shape to the bacterial cell.

In this case, *Curtobacterium* will be rod-shaped and, according to BEDENDO & BELASQUE (2018), with a size of 0.3-0.6 x 0.5-3.0 μm, in yellow or orange and obligatory aerobic colonies. According to OSDAGHI et al. (2020), yellow-colored strains are more predominant and aggressive when compared with other variants. *Curtobacterium falccumfacines* pv. *flaccumfaciens* is the only species of the genus identified in Brazil, where it attacks common beans and soybeans.

Curtobacterium belongs to a group considered a minority among phytopathogenic bacteria, that is of the Gram-positive bacteria. These organisms can be found in the Phylum Actinobacteria and, in the case of the bacteria in question, in the class that have the same name (Figure 8).

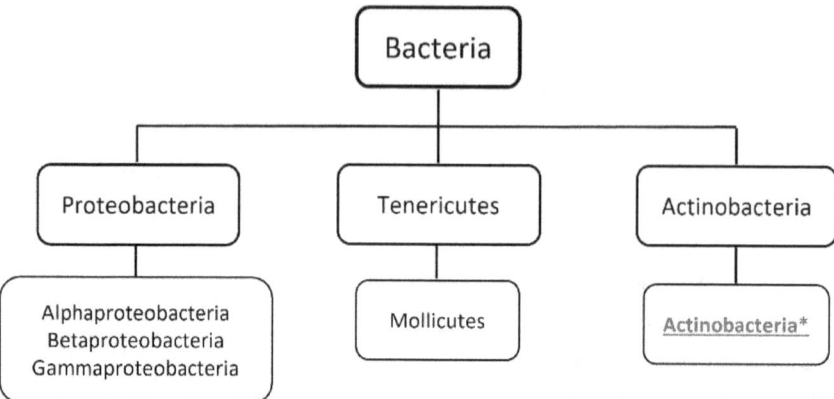

Figure 8. Schematic diagram shows the positioning of the genus *Curtobacterium* within the recent systematics of bacteria. The Phylum Actiniobacteria groups Gram-positive bacteria, while the Phylum Proteobacteria contains Gram-negative bacteria. The Phylum Tenericutes includes bacteria lacking cell wall.

Symptomatology

The bacteria colonize the xylem vessels, obstructing the passage of raw sap, leading to the symptoms that will be discussed below. Therefore, among the symptoms, intercalary flaccidity and chlorosis in the leaflets stand out, which leads to necrotic areas on the leaf blade, surrounded by chlorotic margins (CHEN et al., 2021). As symptoms progress, it is possible to observe wilting of the leaves during the driest day periods, leading to wilting and death of the plant in cases of more severe infections and favorable plant pathogen environmental conditions (HARVESON et al., 2015). Another characteristic symptom comprises in the discoloration of the seed surface in yellow or orange tones and wrinkling (HARDING et al., 2022). Other symptoms are burning, wilting and wrinkling of the leaf margin (PUIA et al., 2021), vascular darkening, dwarfism, stunting and, consequent plant death (WENDLAND et al., 2018).

Curtobacterium wilt can be confused with *Fusarium* wilt, which makes diagnosis difficult. Therefore, isolation of the bacteria and PCR with specific primers for diagnosis are recommended.

GONÇALVES et al. (2021), with a view to confirming the occurrence of bacterial wilt in symptomatic bean plants, used the PCR technique with specific primers CffFOR2 (5'-GTTATGACTGAACTTCACTCC-3') and CffREV4 (5'-GATGTTCCCGTGTGTTCAG-3'), with amplification of a 306 bp DNA fragment. This technique can be added from re-inoculation in susceptible varieties. In this case, pathogenicity assay can be conducted on common bean plants cv. Pérola (GONÇALVES et al., 2021).

Epidemiology
Climate conditions: The survival of bacteria between growing seasons can be influenced by biotic factors such as microbial diversity in the soil and abiotic factors such as humidity, temperature, pH, aeration and soil physical characteristics (LENNON et al., 2012; GONÇALVES et al., 2018). Temperatures above 30°C enables the disease occurrence (WENDLAND et al., 2016). From another aspect, however, when under controlled conditions of temperature at 20°C and humidity between 15 and 22%, *C. flaccumfaciens* pv. *flaccumfaciens* showed longer survival time in soil (SILVA JUNIOR et al., 2012).

Dissemination: *C. flaccumfaciens* pv. *flaccumfaciens* is included in the list of A2 (high risk) quarantine pathogens of the European and Mediterranean Plant Protection Organization and, therefore, in these regions, it is under strict quarantine control (CHEN et al., 2021). The main dissemination form is through seeds (BEDENDO & BELASQUE, 2018). This dissemination mechanism deserves to be highlighted because the bacteria survives in seeds for up to 25 years and the transmission rate from seed to seedling is 100%, affecting the germination (WENDLAND et al., 2016).

Survival: It is interesting to note that most phytopathogenic bacteria do not form spores or resistance structures, which causes the inoculum to survive between growing seasons associated with crop residues or alternative hosts. In addition to the bean crop as volunteer plants, weeds are important alternative hosts for pathogens, being responsible for the survival and dissemination of phytopathogenic bacteria. In this context, results obtained by NASCIMENTO et al. (2020) showed that *Amaranthus viridis* (Amaranthaceae), *Conyza bonariensis, Emilia fosbergii, Galinsoga parviflora, Gnaphalium purpureum* (Asteraceae), *Raphanus sativus, Lepidium virginicum* (Brassicaceae), *Commelina benghalensis* (Commelinaceae), *Ipomoea triloba* (Convolvulaceae), *Cyperus rotundus* (Cyperaceae), *Senna obtusifolia* (Fabaceae), *Digitaria insularis* (Poaceae), *Nicandra physalodes* and *Solanum americanum* (Solanaceae) are potential hosts for *C. flaccumfaciens* pv. *flaccumfaciens*. Therefore, the mentioned authors recommend the eradication of these plants in bean crops, especially in crops with occurrence of *Curtobacterium* wilt.

Control
Exclusion: As an exclusion measure, it is recommended to use certified seeds.

Eradication: Host plants, including volunteer bean plants, must be eradicated. Results obtained by GONÇALVES et al. (2021) highlights the importance of three aspects to be observed in the practice of crop rotation aiming to control *C. flaccumfaciens* pv. *faccumfaciens*: (1) the planting of cultivars with a level of bacterial wilt resistance, (2) the planting of non-host crops of *Curtobacterium* in rotation systems and (3) the area fallowing for a long time period aiming to reduce the population of bacteria in the soil, however used in crop rotation with non-host crops. In this particular research, the mentioned authors found that the cultivars BRS Campeiro, BRS Estilo, IPR Tuiuiú, Tangará and IPR Campos Gerais presented low incidence, low severity and greater productivity. In another evaluation, strains of *C. flaccumfaciens* pv. *faccumfaciens* were recovered from black oat and wheat plants used in the rotation systems. This last component is important to be observed, since the crop rotation when using bacterial hosts compromises the efficiency of this practice. According to NASCIMENTO et al. (2022), the survival of *C. flaccumfaciens* pv. *faccumfaciens* in the soil was negatively influenced by high temperatures and low soil humidity, so that fallow periods that vary from three to four months can reduce the amount of inoculum in soils subjected to these conditions.

Immunization: Adoption of resistant cultivars.

Therapy: There are no products registered for *C. flaccumfaciens* pv. *faccumfaciens* in the database of the Brazilian Ministry of Agriculture, Livestock and Supply.

References

BEDENDO, I.P.; BELASQUE, J. Bactérias fitopatogênicas. In: AMORIM, L.; REZENDE, J.A.M.; BERGAMIN FILHO, A. **Manual de Fitopatologia: princípios e conceitos**. vol.1, 5.Ed. Ouro Fino: Agronômica Ceres, pp.143-160, 2018.

CHEN, G.; KHOJASTEH, M.; TAHERI-DEHKORDI, A.; TAGHAVI, S.M.; RAHIMI, T.; OSDAGHI, E. Complete genome sequencing provides novel insight into the virulence repertories and phylogenetic position of dry beans pathogen *Curtobacterium flaccumfaciens* pv. *flaccumfaciens*. **Phytopathology**, v.111, p.268-280. 2021. https://doi.org/10.1094/PHYTO-06-20-0243-R

GONÇALVES, R.M.; SILVA JÚNIOR, T.A.F.; SOMAN, J.M.; SILVA, J.C.; MARINGONI, A.C. Effect of crop rotation on common bean cultivars against bacterial wilt caused by *Curtobacterium flaccumfaciens* pv. *flaccumfaciens*. **European Journal of Plant Pathology**, v.159, p.485-493, 2021. https://doi.org/10.1007/s10658-020-02176-6

GONÇALVES, R.M.; SOMAN, J.M.; KRAUSE-SAKATE, R.; PASSOS, J.R.S.; SILVA JÚNIOR, T.A.F.; MARINGONI, A.C. Survival of *Curtobacterium flaccumfaciens* pv. *flaccumfaciens* in the soil under Brazilian conditions. **European Journal of Plant Pathology**, v.152, p.213-223, 2018. https://doi.org/10.1007/s10658-018-1466-z

HARDING, M.W.; MARQUES, L.L.R.; ALLAN, N.; OLSON,

M.E.; BUZIAK, B.; NADWORNY, P.; OMAR, A.; HOWARD, R.J.; FENG, J. Bactericidal efficacy of oxidized silver against biofilms formed by *Curtobacterium flaccumfaciens* pv. *flaccumfaciens*. **The Plant Pathology Journal,** v.38, n.4, p.334-344, 2022. https://doi.org/10.5423/PPJ.OA.04.2022.0055

HARVESON, R.M.; SCHWARTZ, H.F.; URREA, C.A.; YONTS, C.D. Bacterial wilt of dry-edible beans in the central high plains of the U.S.: Past, Presente and Future. **Plant Disease,** v.99, n.12, p.1665-1677, 2015. https://doi.org/10.1094/PDIS-03-15-0299-FE

LENNON, J.T.; AANDERUD, Z.T.; LEHMKUHL, B.K.; SCHOOLMASTER, D.R. Mapping the niche space of soil microorganisms using taxonomy and traits. **Ecology,** v.93, n.8, p.1867-1879, 2012. https://doi.org/10.1890/11-1745.1.

NASCIMENTO, D.M.; OLIVEIRA, L.R.; MELO, L.L.; RIBEIRO-JUNIOR, M.R.; SILVA, J.C.; SOMAN, J.M.; SARTORI, M.M.P.; SILVA JÚNIOR, T.A.F.; MARINGONI, A.C. Survival of *Curtobacterium flaccumfaciens* pv. *flaccumfaciens* from soybean and common bean in soil. **European Journal of Plant Pathology,** v.162, p.971-979, 2022. https://doi.org/10.1007/s10658-021-02451-0

NASCIMENTO, D.M.; OLIVEIRA, L.R.; MELO, L.L.; SILVA, J.C.; SOMAN, J.M.; GIROTTO, K.T.; EBURNEO, R.P.; RIBEIRO-JUNIOR, M.R.; SARTORI, M.M.P.; SILVA JÚNIOR, T.A.F.; MARINGONI, A.C. Survival of *Curtobacterium flaccumfaciens* pv. *flaccumfaciens* in weeds. **Plant Pathology**, v.69, p.1357-1367, 2020. https://doi.org/10.1111/ppa.13206

OSDAGHI, E.; YOUNG, A.J.; HARVESON, R.M. Bacterial wilt of dry beans caused by *Curtobacterium flaccumfaciens* pv. *flaccumfaciens*: A new threat from an old enemy. **Molecular Plant Pathology,** v.21, p.605-621, 2020. https://doi.org/10.1111/mpp.12926

PUIA, J.D.; FERREIRA, M.G.D.B.; HOSHINO, A.T.; BORSATO, L.C.; CANTERI, M.G.; VIGO, S.C. Occurrence of *Curtobacterium flaccumfaciens* pv. *flaccumfaciens* in the state of Paraná and its pathogenicity in beans. **European Journal of Plant Pathology,** v.159, p.627-636, 2012. https://doi.org/10.1007/s10658-020-02193-5

SILVA JÚNIOR, T.A.F.; NEGRÃO, D.R.; ITAKO, A.T.; SOMAN, J.M.; MARINGONI, A.C. Survival of *Curtobacterium flaccumfaciens* pv. *flaccumfaciens* in soil and bean crop debris. **Journal of Plant Pathology,** v.94, n.2, p.331-337, 2012. https://doi.org/10.4454/JPP.FA.2012.025

WENDLAND, A.; LOBO JUNIOR, M.; FARIA, J.C. **Manual de identificação das principais doenças do feijoeiro-comum.** Empresa Brasileira de Pesquisa Agropecuária, Embrapa Arroz e Feijão, Ministério da Agricultura, Pecuária e Abastecimento. Brasília: Embrapa, 2018. 49p.

WENDLAND, A.; MOREIRA, A.S.; BIANCHINI, A.; GIAMPAN, J.S.; LOBO JUNIOR, M. Doenças do Feijoeiro. In: AMORIM, L.; REZENDE. J.A.M.; BERGAMIN FILHO, A.; CAMARGO, L.E.A. **Manual de Fitopatologia: Doenças das plantas cultivadas.** vol.2, 5.Ed. Ouro Fino: Agronômica Ceres, pp.383-396, 2016.

6. Anthracnose - *Colletotrichum lindemuthianum*

Kingdom: Fungi
Phylum: Ascomycota
Order: Glomerellales
Family: Glomerellaceae

Etiology
There are 68 orders in the Phylum Ascomycota, of which 18 are important within phytopathology (MASSOLA JÚNIOR, 2018). Another point to be considered is the double nomenclature that exists in the case of fungi from the Phylum Ascomycota. In this sense, attention must be paid to the order and family in which the fungus is found when in its teleomorphic phase (*Glomerella cingulata* f. sp *phaseoli*), to the detriment of its anamorph (*Colletotrichum lindemuthianum*).

The Glomerellaceae family is the most important family for phytopathology within the Glomerellales order. Among the notable morphological characters for the order Glomerellales and family Glomerellaceae, stands out the sexual fruiting body of the Perithecium type (SUTTON & SHANE, 1983), without stroma, unitunicate asci, which produce unicellular, hyaline, ellipsoid, straight or slightly curved ascospores (MASSOLA JÚNIOR, 2018). The anamorph *Colletotrichum* is the causal agent of a wide range of hosts of agronomic and economic importance (SILVA et al., 2020), such as mango (ASSUNAÇÃO et al., 2018) and annual crops, such as beans (NABI et al. 2022) and soybeans (BOUFLEUR et al., 2021).

The species *Colletotrichum lindemuthianum* presents great genetic, morphological and physiological variability (SOUZA et al., 2007; NABI et al., 2022). Important phenotypic characters used in the taxonomy of the genus *Colletotrichum* include the dimensions of non-germinated conidia and the analysis of the presence of septa in germinated conidia (O'CONNELL et al., 1992).

In this sense, the main characters of the pathogen are the hyaline, cylindrical, aseptate conidia, measuring 3.5-8.3 x 11.5-20.7 μm (PINTO et al., 2012). In addition, acervuli are always found, which have dark and clearly evident arrows. Figure 9 outlines the positioning of the fungus *C. lindemuthianum* on the current taxonomy map of the Kingdom Fungi.

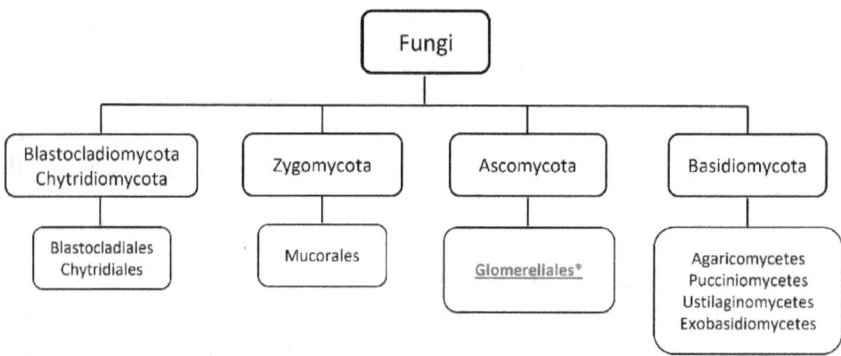

Figure 9. Schematic diagram shows the positioning of the fungus *Colletotrichum lindemuthianum* within the order Glomerellales, of the Fungi kingdom.

Simptomatology

Although infection can occur on both leaf sides, the initial signs of infection usually appear on the abaxial side along the veins, which show a reddish discoloration. Later, this discoloration also appears on the adaxial face (MOHAMMED, 2013; NABI et al., 2022). The most typical symptom is found in the pods, where deep circular lesions are observed (Figure 10), with a brown or reddish border, and may even contain signs of the pathogen such as the presence of a reddish spores mass (PADDER et al., 2017; NABI et al., 2022).

Figure 10. Common bean anthracnose: typical symptom on the pods, where depressed circular lesions are observed, with a brown or reddish border.

Epidemiology
Climate conditions: Cloudy days with moderate temperatures and high humidity are favorable for the disease (NABI et al., 2022).

Dissemination: In the field, cultural remains contribute to dissemination and survival, while seeds play an important role not only in survival, but also in dissemination over long distances.

Survival: In addition to surviving on seeds, the pathogen survives in crop remains and alternative hosts, such as *Phaseolus lunatus, P. coccieus, P. acutitolius, Vigna radiate, Vigna unguiculata, Dolichos bitloris, Vicia faba, Glycine max, Pisum sativum* and *Vigna mungo* (MOHAMMED, 2013).

Control
Exclusion: Seeds obtained from fields previously infected by

anthracnose should not be used (BUSH, 2009). Therefore, it is recommended to use of healthy and certified seeds, obtained from production fields whose conditions are not favorable for anthracnose.

Eradication: Crop remains should be removed after harvest aiming to reduce winter survival. Furthermore, a two-year crop rotation, using non-host species (cereals and nightshades), is recommended to minimize the fungus survival (MOHAMMED, 2013).

Regulation: BUSH (2009) also reinforces the need to ensure adequate spacing between plants, especially in fields intended for seed production. The practice of irrigation deserves attention, as it favors the release of spore masses on the foliage by the water splashes.

Immunization: Resistant varieties can be used, despite the great genetic variability and occurrence of more than 182 races of *C. lindemuthianum* around the world (PADDER et al., 2017).

Therapy: When it comes to therapy, it is important to mention the use of fungicides to spray the aerial part of plants, as well as to seeds treatment (MOHAMMED, 2013). In Brazil, there are a good number of registered products. When carrying out a brief consultation on the AGROFIT website of the Ministry of Agriculture, Livestock and Supply (MAPA) (https://agrofit.agricultura.gov.br/agrofit_cons/principal_agrofit_co ns) for products registered for *Colletotrichum lindemuthianum*, there are 132 fungicides cataloged in the section of products formulated for aerial/terrestrial application and 22 fungicides for seed treatment (AGROFIT, 2022).

References
AGROFIT. **Sistemas de agrotóxicos fitossanitários**. Ministério da Agricultura, Pecuária e Abastecimento. Disponível em: <http://extranet.agricultura.gov.br/agrofit_cons/principal_agrofit_c ons> Acesso em 12/08/2022.

ASSUNÇÃO, M.C.; AMARAL, A.G.G.; LINS, F.J.A. Efeito da temperatura e de embalagens sobre a antracnose em frutos de Manga cv. Tommy Atkins. **Ciência Agrícola,** v.16, n.3, p.35-42, 2018. https://doi.org/10.28998/rca.v16i3.3490

BOUFLEUR, T.R.; MASSOLA JÚNIOR, N.S.; TIKAMI, Í.; SUKNO, S.A.; THON, M.R.; BARONCELLI, R. Identification and Comparison of Colletotrichum Secreted Effector Candidates Reveal Two Independent Lineages Pathogenic to Soybean. **Pathogens,** v.10, e-1520, 2021. https://doi.org/10.3390/pathogens10111520

MASSOLA JÚNIOR, N.S. Fungos fitopatogênicos. In: AMORIM, L.; REZENDE, J.A.M.; BERGAMIN FILHO, A. **Manual de Fitopatologia: princípios e conceitos.** vol.1, 5.Ed. Ouro Fino: Agronômica Ceres, pp.107-142, 2018.

MOHAMMED, A. An overview of distribution, biology and the management of common bean anthracnose. **Journal of Plant Pathology & Microbiology,** v.4, n.8, p.193, 2013. https://doi.org/10.4172/2157-7471.1000193

NABI, A.; LATEEF, I.; NISA, Q.; BANOO, A.; RASOOL, R.S.; SHAH, M.D.; AHMAD, M.; PADDER, B.A. *Phaseolus vulgaris - Colletotrichum lindemuthianum* pathosystem in the post-genomic era: an update. **Current Microbiology,** v.79:36, 2022. https://doi.org/10.1007/s00284-021-02711-6

O'CONNELL, R.J.; NASH, C.; BAILEY, J.A. Lectin citochesmitry: a new approach to understanding cell differentiation, pathogenesis and taxonomy in *Colletotrichum*. In: BAYLEY, J.A.; JEGER, M.J. *Colletotrichum*: **Biology, Pathology and Control,** Wallingford: CAB International, pp.67-87, 1992.

PADDER, B.A.; SHARMA, P.N.; AWALE, H.E.; KELLY, J.D. *Colletotrichum lindemuthianum*, the causal agent of bean anthracnose. **Journal of Plant Pathology,** v.99, n.2, p.317-330,

2017. http://dx.doi.org/10.4454/jpp.v99i2.3867

PINTO, J.M.A.; PEREIRA, R.; MOTA, S.F.; ISHIKAWA, F.H.; SOUZA, E.A. Investigating phenotypic variability in *Colletotrichum lindemuthianum* populations. **Phytopathology**, v.102, p.490-497, 2012. https://doi.org/10.1094/PHYTO-06-11-0179

SILVA, L.L.; MORENO, H.L.A.; CORREIA, H.L.N.; SANTANA, M.F.; QUEIROZ, M.V. Colletotrichum: species complexes, lifestyle, and peculiarities of some sources of genetic variability. **Applied Microbiology and Biotechnology**, v.104, p.1891-1904, 2020. https://doi.org/10.1007/s00253-020-10363-y

SOUZA, B.O.; SOUZA, E.A.; MENDES-COSTA, M.C. Determinação da variabilidade em isolados de *Colletotrichum lindemuthianum* por meio de marcadores morfológicos e culturais. **Ciência e Agrotecnologia**, v.31, p.1000-1006, 2007. https://doi.org/10.1590/S1413-70542007000400009

SUTTON, T.B., SHANE, W.W. Epidemiology of the perfect stage of *Glomerella cingulata* on apples. **Phytopathology**, v.73, p.1179-1183, 1983. https://doi.org/10.1094/Phyto-73-1179

7. Angular leaf spot - *Pseudocercospora griseola*

Kingdom: Fungi
Phylum: Ascomycota
Order: Capnodiales
Family: Mycosphaerellaceae

Etiology
When it comes to the order Capnodiales, especially the Mycospaerellaceae family, which is the main one in the order, there is the genus *Mycosphaerella*, which is a teleomorph of a wide range of anamorphic species of agronomic importance, such as *Pseudocercospora* in beans (PÁDUA et al., 2022) and *Cercospora* and *Septoria* in soybeans (GODOY et al, 2016).

The classification based on the morphology of the sexual reproduction structure has been revised using molecular techniques. LUMBSCH & HUHNDORF (2007) had already addressed the problematic circumscription of the Loculoascomycetes class, as well as the classification and proposition of orders within this class. In this sense, we emphasize that for the phylum Ascomycota, especially in this work, directions will be made to the orders of economic importance occurring within the phylum, and there is no proposal for classification for a classes level within the phylum Ascomycota.

Pseudocercospora griseola produces conidia in bundles of dark-colored conidiophores (synnemata) on the leaves abaxial surface.

Conidia are obclavate-cylindrical, largely subfusiform, while short conidia are sometimes ellipsoid-ovoid to short cylindrical, straight to curved measuring 20.0-85.0 µm long by 4.0-9.0 µm central wide, septate and measuring 1.5-3.0 µm in basal width (CROUS et al., 2006). The level of variability between and within *Pseudocrecospora griseola* populations is considerably high (LIEBENBERG & PRETORIUS, 1997). In addition to the pathogenic variability already reported in Brazil (SILVA et al.,

2008), Argentina (STENGLEIN et al., 2006) and Turkey (CANPOLAT & MADEN, 2020), there is also variability in the morphology of this fungus. The data regarding the morphological characterization of *P. griseola*, even within the measurement ranges proposed by CROUS et al. (2006), are variable. As an example, we can cite the work of LIBRELON et al. (2022), who, after evaluating 125 isolates from different locations in the state of Minas Gerais, Brazil, found measurements of 33.6-43.4 μm in length by 5.2-10.9 in width.

In the specific case of common beans, the most important member of the Capnodiales order is the fungus *Pseudocercospora griseola*, the causal agent of bean angular spot (Figure 11). The causal agent was previously known as *Phaeoisariopsis griseola*, until CROUS et al. (2006) reassessed the taxonomic status for this species. Analysis of SSU nrDNA (Small subunit nuclear ribosomal deoxyribonucleic acid) sequences revealed that the genus *Phaeoisariopsis* was indistinguishable from other anamorphic Hyphomycetes genera associated with *Mycosphaerella*, represented by *Pseudocrecospora* and *Stigmina*. Thus, a new combination was proposed in the genus *Pseudocrecospora* as a name to be conserved and later adopted.

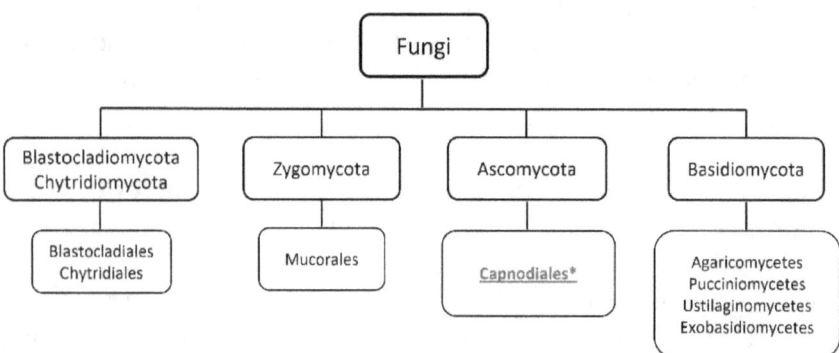

Figure 11. Schematic diagram shows the positioning of the fungus *Pseudocercospora griseola* within the order Capnodiales, of the Fungi kingdom.

Symptomatology
The symptom on the pods consists of reddish-brown lesions with a

circular to elliptical shape, while on the leaves the beginning is marked by the presence of small brown to gray spots that become necrotic and angular (Figure 12), delimited by the veins (CROUS et al., 2006; REZENE et al., 2018). That is, the disease receives its name due to the fact that the resulting lesion design forms angles (90° and 60°) depending on the angle of insertion of one vein in relation to the other responsible for the formation of the lesion border area on the leaf tissue (Figure 13). The lesions are more visible in the final stages of the crop, where they may present a yellowish halo, coalescence and defoliation.

Figure 12. Common bean angular spot: necrotic and angular brown spots, that is, delimited by veins, found on plants in the crop final stages.

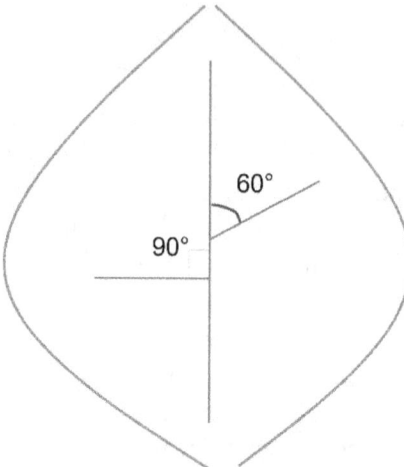

Figure 13. Schematic shows formation of angles (90° and 60°) between the border veins for the formation of an angular bean spot lesion on the leaf tissue.

Epidemiology

Climate conditions: Optimal conditions for infection include humidity and a temperature of 24°C. Under these conditions, conidia can germinate in about 3 hours, penetrating the leaves through the stomata within 2 days. Then, the entire tissue is colonized by the pathogen after 3 to 7 days (LIEBENBERG & PRETORIUS, 1997; LIBRELON et al., 2022).

Dissemination: The fungus *Pseudocercospora griseola* can be transmitted through seeds. However, the most frequent source of primary inoculum under natural conditions is the presence of cultural remains that contain the pathogen (NAY et al., 2019). Conidia are also spread by the wind, in water droplets and agricultural implements (LIBRELON et al., 2022).

Survival: A wide range of alternative hosts are observed for this pathogen: *Lablab niger, L. purpureus, Lathyryus odoratus, Macroptilium arthropurpureum, Phaseolus acutifolius, P. aureus, P. coccineus, P. lunatus, P. pubescens, P. vulgaris, Vigna angularis , V. mungo, V. radiata, V. sinensis* and *V. unguiculata* (CROUS et al., 2006).

Control
Eradication: Most varieties present moderate resistance, a reality that highlights the need for the joint use of other control measures such as the elimination of alternative hosts and cultural remains.

Immunization: The control of the common bean angular leaf spot through the use of resistant cultivars is difficult by the extensive virulence diversity of *Pseudocercospora griseola* and the recurrent appearance of new virulent races. When it comes to common bean genetic improvement aiming to obtain resistance to angular leaf spot, five main loci (Phg-1, Phg-2, Phg-3, Phg-4 and Phg-5) were named and markers strongly linked to these loci have been reported. The bean reference genome and new sequencing technologies have allowed the development of molecular markers closely linked to Phg loci. In this context, NAY et al (2019) published a review to serve as a reference for future studies on resistance mapping and selection of resistance loci. Obtaining a bean cultivar with resistance to angular leaf spot is a challenging topic. PÁDUA et al. (2022), after evaluating the reaction of 416 germplasm accessions to *P. griseola* (race 63-63, the most important and aggressive race) under greenhouse conditions, they obtained 24.5% resistant accessions, of which little more half were from the Carioca group.

Therapy: Due to the presented circumstances, the use of fungicides is a recurring practice in common bean cultivation. According to LIBRELON et al. (2022), in Brazil, 151 commercial fungicides are registered for controlling angular leaf spot, of which the following chemical groups can be mentioned: benzimidazole, dithiocarbamate, inorganic groups (copper and tin), strobilurins, triazoles, isophthalonitrile and some mixtures.

References
CANPOLAT, S.; MADEN, S. Reactions of some common bean cultivars grown in Turkey against some isolates of angular leaf spot disease, caused by *Pseudocercospora griseola* (Sacc.) Crous & U. Braun. ***Plant Protection Bulletin***, *v.60, n.2, p.45-54, 2020.*

https://doi.org/10.16955/bitkorb.630968

CROUS, P.W.; LIEBENBERG, M.M.; BRAUN, U.; GROENEWALD, J.Z. Re-evaluating the taxonomic status of *Phaeoisariopsis griseola*, the causal agent of angular leaf spot of bean. **Studies in Mycology,** v.55, p.163-173, 2006. https://doi.org/10.3114/sim.55.1.163

GODOY, C.V.; ALMEIDA, A.M.R.; COSTAMILAN, L.M.; MEYER, M.C.; DIAS, W.P.; SEIXAS, C.D.S.; SOARES, R.M.; HENNING, A.A.; YORINORI, J.T.; FERREIRA, L.P.; SILVA, J.F.V. Doenças da Soja. In: AMORIM, L.; REZENDE. J.A.M.; BERGAMIN FILHO, A.; CAMARGO, L.E.A. **Manual de Fitopatologia: Doenças das plantas cultivadas.** vol. 2, 5. Ed. Ouro Fino: Agronômica Ceres, pp. 657-675. 2016.

LIBRELON, S.S.; PEREIRA, F.A.C.; PÁDUA, P.F.; PEREIRA, N.B.M.; GOMES, L.B.W.; PEREIRA, R.; PEREIRA, L.F.; POZZA, E.A.; SOUZA, E.A. *Pseudocercospora griseola*, the causal agent of common bean angular leaf spot: Strain characterization and sensitivity to fungicides. **Plant Pathology,** v.72, n.6, p.1-9, 2022. https://doi.org/10.1111/ppa.13556

LIEBENBERG, M.M.S.; PRETORIUS, Z.A. A review of angular leaf spot of common bean (*Phaseolus vulgaris* L.). **African Plant Protection,** v.3, n.2, p.81-106, 1997.

LUMBSCH, H.T.; HUHNDORF, S.M. Whatever happened to the pyrenomycetes and loculoascomycetes? **Mycological Research,** v.3, p.1064-1074, 2007. https://doi.org/10.1016/j.mycres.2007.04.004

NAY, M.M.; SOUZA, T.L.P.O.; RAATZ, B.; MUKANKUSI, C.M.; GONÇALVES-VIDIGAL, M.C.; ABREU, A.F.B.; MELO, L.C.; PASTOR-CORRALES, M.A. A review of angular leaf spot resistance in common bean. **Crop Science,** v.59, p.1376-1391, 2019. https://doi.org/10.2135/cropsci2018.09.0596

PÁDUA, P.F.; BARCELOS, Q.L.; PEREIRA, F.A.C.; GOMES, L.B.W.; SOUZA, E.A. Identification of sources of resistance to race 63-63 of *Pseudocercospora griseola* in common bean lines. **Crop Breeding and Applied Biotechnology**, v.22, n.1, e36982215, 2022. https://doi.org/10.1590/1984-70332022v22n1a05

REZENE, Y.; TESFAYE, K.; CLARE, M.; GEPTS, P. Pathotypes characterization and virulence diversity of *Pseudocercospora griseola* the causal agent of angular leaf spot disease collected from major common bean *(Phaseolus vulgaris L.)* growing areas of Ethiopia. **Journal of Plant Pathology and Microbiology**, v.9, n.8, 445, 2018. https://doi.org/10.4172/2157-7471.1000445

SILVA, K.J.D.E.; SOUZA, E.A.; SARTORATO, A.; SOUZA FREIRE, C.N. Pathogenic variability of isolates of *Pseudocercospora griseola*, the cause of common bean angular leaf spot, and its implications for resistance breeding. **Journal of Phytopathology**, v.156, n.10, p.602-606, 2008. https://doi.org/10.1111/j.1439-0434.2008.01413.x

STENGLEIN, S.A.; BALATTI, P.A. First report of angular leaf spot caused by *Phaeoisariopsis griseola* on *Phaseolus coccineus* in Argentina. **Plant Disease**, v.90, n.2, p.248, 2006. https://doi.org/10.1094/PD-90-0248B

8. White mold - *Sclerotinia sclerotiorum*

Kingdom: Fungi
Phylum: Ascomycota
Order: Helotiales
Family: Sclerotiniaceae

Etiology
The order Helotiales is known for having fungi that produce sclerotia and, from these, the Helotiales typical structure apothecia. Sclerotia comprise a resistance structure obtained from the hyphae densification (WILLETS, 1971). These structures help the fungus survive on the soil under adverse conditions such as low temperatures, desiccation, microbial attack and prolonged absence of a plant host (SMITH et al., 2015).

The fungus *Sclerotinia sclerotiorum* is a homothallic and haploid fungus that exhibits sexual and asexual reproduction through resistance structures known as sclerotia (Figure 14). In asexual reproduction, sclerotia present myceliogenic germination (emission of vegetative mycelium), while in a sexual reproduction, sclerotia will present carpogenic germination, producing apothecia (sexual fruiting body), which has ascus containing ascospores (ABÁN et al., 2021).

Literature reports on the number of apothecia produced by sclerotia are very variable. To evaluate this aspect, the conditions need to be previously determined, as well as the material to be subjected to analysis. The study conducted in laboratory by VENTUROSO et al. (2014), not only carried out a separation between different situations (sclerotium position: superficial x buried in the soil) to evaluate the production of the number of apothecia per sclerotia, but also separated the evaluated sclerotia according to weight, which were classified into six classes: (C1) sclerotia with a mass of less than 0.01 g, (C2) 0.01<0.02 g, (C3) 0.02<0.03 g, (C4) 0.03<0.04 g, (C5) 0.04<0.05 g and (C6) 0.05<0.06 g. After the experiments, the authors concluded that sclerotia with greater mass and located on the soil surface had a greater number of apothecia per sclerotium. Another interesting data reports that the apothecium can produce up to 7.6 x 10⁵ ascospores in the air for one hour over a 20 days-period (CLARKSON et al., 2003).

When obtained from *in vitro* production under 12 hours of light, each sclerotium has an average mass of 16.9 mg (PEREIRA et al., 2016). Most Helotiales fungi have tiny apothecia, generally less than 2 mm in diameter (HOSOYA, 2021). However, in the case of *S. sclerotiorum*, these present 4-10 mm in diameter (WENDLAND et al., 2016). Additionally, apothecia may be sessile, dark to shiny in color, and superficial or eruptive through the host plant. The general shape of the apothecia is cupulate-discoid, funnel-shaped or clavate (KORF, 1973).

Ascospores are ejected from asci located in the hymenium of the apothecium. Each asci has 8 sexual ascospores, obtained after plasmogamy of compatible hyphae, karyogamy and meiosis of the formed dikaryon. The ascospores are hyaline, binucleate, ellipsoid and measure 4.0-6.0 x 9.0-14.0 μm (KOHN, 1979).

Figure 14. Common bean white mold: sclerotia of *Sclerotinia*

sclerotiorum, the structures that provide resistance to the fungus.

Normally, between 8 and 10 families are appointed to this order. However, in an extensive review, HOSOYA (2021) reports the existence of 25 families, among which Sclerotiniaceae is one of the most important (Figure 15).

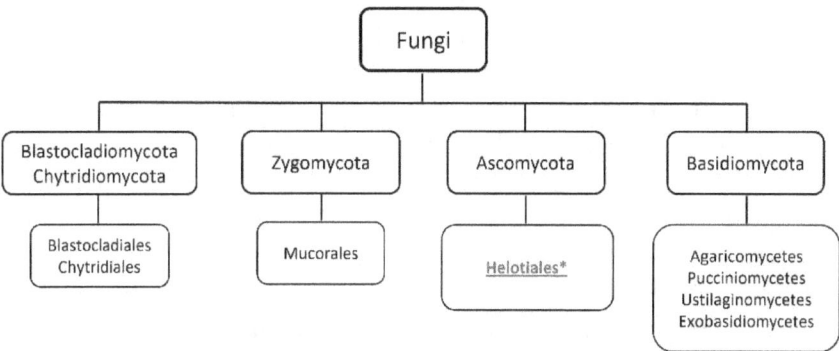

Figure 15. Schematic diagram shows the positioning of the fungus *Sclerotinia sclerotiorum* within the order Helotiales, of the Fungi kingdom.

Symptomatology
White mold is an easy-to-detect disease, characterized by the occurrence of white, cottony and abundant mycelium on the plant tissue of the attacked organs, accompanied by a tissue waterlogging. Subsequently, the lesions dry and the formation of dark, irregularly shaped sclerotia occurs, inside and outside stems and pods (STEADMAN & BOLAND, 2005; BOLTON et al., 2006). With only mycelogenic germination occurring, the plant aerial part can also be affected when they come into contact with the contaminated soil, which appears exhibiting mycelial growth of the fungus on the surface, as observed by MITSUEDA & CHARCHAR (1994). Direct attack by *S. sclerotiorum* mycelium on leaves is not common, except when inoculated. Even in this situation, occurs waterlogging and colonization of the inoculated leaf tissue.

Epidemiology
Climate conditions: The production of apothecia depends on

environmental conditions such as moist soil and mild temperatures (16 to 24°C) (MILA & YANG, 2008). White mold epidemics coincide with the bloom. This occurs because senescent flower parts serve as the primary nutrient source as they fall onto leaves, petioles or stems. Furthermore, the crop flowering occurs at the time of canopy closure and consequently, nutrient sources are available during a period in which environmental conditions are more favorable for the pathogen growth (BOLTON et al., 2006).

Dissemination: There are several ways in which the disease can be spread: infected seeds, sclerotia mixed with seeds, infested soil, irrigation water and ascospores carried by the wind (STEADMAN & BOLAND, 2005).

Survival: Sclerotia introduced or resulting from a white mold epidemic will contribute to the long-term area infestation. This occurs because according to ABÁN et al., (2020), these resistance structures of *S. scleotiorum* can survive in the soil for up to five years or longer, making this disease difficult to control.

Control
Evasion: It is recommended to avoid contaminated areas. To get an idea of the extent of contamination in an area, CARVALHO et al. (2015) measured that an infested area without application of control measures may contain between 12.5 and 105.5 scleroria by m^2, that is, a population that was sufficient to cause 28.5 and 58.7% severity in common bean cv. Perola, respectively. In addition to area evasion, the evasion in time, that is, the use of early cultivars demonstrated significant results when in field experiments, especially the cultivars BRS Radiante and Jalo Precoce (LOBO JÚNIOR et al., 2009).

Exclusion: The ideal control principle comprises in the exclusion of the pathogen, that is, maintaining the condition of the planting area unharmed. As the introduction of the pathogen often occurs through infected seeds or sclerotia mixed with the seeds, it is recommended to use certified seeds in addition to their treatment. In addition to seeds, the pathogen may enter via equipment used in infested areas, which suggests that these be disinfested before entering uninfested areas.

Eradication: After introduction into the planting site, one of the alternatives is to reduce the pathogen population by destroying the sclerotia. In this sense, one of the most researched measures consists of the periodic application of the hyperparasitic fungus *Trichoderma* spp. This biocontrol agent has the ability to colonize and destroy sclerotia previously existing in the soil. It is important to emphasize that the biocontrol of diseases promoted by *Trichoderma* consists of a complex process that can occur through antibiosis, competition for nutrients, hyperparasitism, among others (HARMAN et al., 2004). Another eradication measure consists of treating seeds with chemical fungicide (Carboxin+Thiram), which showed good results, as demonstrated by CARVALHO et al. (2011).

Regulation: Among cultural control measures, an interesting option is the use of straw in the soil, since this inhibits the formation of apothecia and, consequently, ascospores (FERRAZ et al., 1999; FERGUSON et al., 2001). Furthermore, the formation of straw can become essential for the employ of biological control in annual crops, since biocontrol activity is greater when in an environment formed by straw (GORGEN et al., 2009). Another regulation measure consists in upright cultivars (erect plants), which showed less white mold severity compared to those with prostrate growth, when evaluated in a sprinkler-irrigated experiment, without the fungicides application (LOBO JÚNIOR et al., 2009).

Immunization: Commercial common bean cultivars resistant to white mold are not available, but materials with partial resistance have been identified in controlled conditions and field trials with the bean crop (ABÁN et al., 2020). Physiological resistance and plant architectural characteristics, such as the erect plants, allowing the reduction of white mold and are used for the development of cultivars in common bean genetic improvement programs (LIMA et al., 2019; ABÁN et al., 2020).

Therapy: The application of fungicides has been used to manage white mold on beans, but areas with high infestations require many applications, which increases production costs (CARVALHO et al., 2015).

References

ABÁN, C.L.; TABOADA, G.; SPEDALETTI, Y.; MAITA, E.; GALVÁN, M.Z. Population structure of the fungus *Sclerotinia sclerotiorum* in common bean fields of Argentina. **European Journal of Plant Pathology,** v.160, p.841-853, 2021. https://doi.org/10.1007/s10658-021-02288-7

ABÁN, C.L.; TABOADA, G.M.; CASALDERREY, N.B.; MAGGIO, M.E.; CHOCOBAR, M.O.; SPEDALETTI, Y.A.; GONZALEZ. M.A.A.; VIZGARRA, O.N.; GALVÁN, M.Z. Screening common bean germplasm for resistance to genetically diverse *Sclerotinia sclerotiorum* isolates from Argentina. **Acta Scientiarum Agronomy,** v.42, e42786, 2020. https://doi.org/10.4025/actasciagron.v42i1.42786

BOLTON, M.D.; THOMMA, B.P.H.J.; NELSON, B.D. *Sclerotinia sclerotiorum* (Lib.) de Bary: biology and molecular traits of a cosmopolitan pathogen. **Molecular Plant Pathology,** v.7, n.1, p.1-16, 2006. https://doi.org/10.1111/j.1364-3703.2005.00316.x

CARVALHO, D.D.C.; GERALDINE, A.M.; LOBO JUNIOR, M.; MELLO, S.C.M. Biological control of white mold by *Trichoderma harzianum* in common bean under field conditions. **Pesquisa Agropecuária Brasileira,** v.50, n.12, p.1220-1224, 2015. https://doi.org/10.1590/S0100-204X2015001200012

CARVALHO, D.D.C.; MELLO, S.C.M.; LOBO JÚNIOR, M.; GERALDINE, A.M. Biocontrol of seed pathogens and growth promotion of common bean seedlings by Trichoderma harzianum. **Pesquisa Agropecuária Brasileira,** v.46, n.8, p.822-828, 2011. https://doi.org/10.1590/S0100-204X2011000800006

CLARKSON, J.P.; STAVELEY, J.; PHELPS, K.; YOUNG, C.S.; WHIPPS, J.M. Ascospore release and survival in *Sclerotinia sclerotiorum*. **Mycological Research,** v.107, n.2, p.213-222, 2003. https://doi.org/10.1017/S0953756203007159

FERGUSON, L.M.; SHEW, B.B. Wheat straw mulch and its impacts on three soilborne pathogens of peanut in microplots. **Plant Disease,** v.85, p.661-667, 2001. https://doi.org/10.1094/PDIS.2001.85.6.661

FERRAZ, L.C.L.; CAFÉ FILHO, A.C.; NASSER, L.C.B.; AZEVEDO, J.A. Effects of soil moisture, organic matter and grass mulching on the carpogenic germination of sclerotia and infection of bean by *Sclerotinia sclerotiorum*. **Plant Pathology,** v.48, p.77-82, 1999. https://doi.org/10.1046/j.1365-3059.1999.00316.x

GORGEN, C.A.; SILVEIRA NETO, A.N.; CARNEIRO, L.C.; RAGAGNIN, V.; LOBO JUNIOR, M. Controle do mofo-branco com palhada e *Trichoderma harzianum* 1306 em soja. **Pesquisa Agropecuária Brasileira,** v.44, n.12, p.1583-1590, 2009. https://doi.org/10.1590/S0100-204X2009001200004

HARMAN, G.E.; HOWELL, C.R.; VITERBO, A.; CHET, I.; LORITO, M. *Trichoderma* species - opportunistic, avirulent plant symbionts. **Nature Reviews Microbiology,** v.2, p.43-56, 2004. https://doi.org/10.1038/nrmicro797

HOSOYA, T. Systematics, ecology, and application of *Helotiales*: Recent progress and future perspectives for research with special emphasis on activities within Japan. **Mycoscience,** v.62, p. 1-9, 2021. https://doi.org/10.47371/mycosci.2020.05.002

KOHN, L.M. A monographic revision of the genus *Sclerotinia*. **Mycotaxon,** v.9, p.365-444, 1979.

KORF, R.P. Discomycetes and Tuberales. In: AINSWORTH, G.C.; SPARROW, F.K.; SUSSMAN, A.S. **The Fungi: an advanced**

treatise. New York: Academic Press. pp.249-319, 1973.

LIMA, R.C.; TEIXEIRA, P.H.; SOUSA, L.R.V.; RODRIGUES, L.B.; CARNEIRO, J.E.S.; LEHNER, M.S.; PAULA JUNIOR, T.J.; VIEIRA, R.F. Integration of partial resistance, plant density and use of fungicide for management of white mould in common bean. **Plant Pathology**, v.68, p.481-491, 2019. https://doi.org/10.1111/ppa.12973

LOBO JUNIOR, M.; GERALDINE, A.M.; CARVALHO, D.D.C.; COBUCCI, T. **Uso de Cultivares de Feijão Comum com Arquitetura Ereta e Ciclo Precoce para Escape do Mofo Branco (*Sclerotinia sclerotiorum*).** Comunicado técnico 182. Santo Antônio de Goiás: Empresa Brasileira de Pesquisa Agropecuária. 2009, 4p.

MILA, A.L.; YANG, X.B. Effects of fluctuating soil temperature and water potential on sclerotia germination and apothecial production of *Sclerotinia sclerotiorum*. **Plant Disease,** v.92, p.78-82, 2008. https://doi.org/10.1094/PDIS-92-1-0078

MITSUEDA, T.; CHARCHAR, M.J.D.A. Modo de ocorrência do mofo-branco (*Sclerotinia sclerotiorum*) em feijoeiro irrigado na região dos cerrados. In: Centro de Pesquisa Agropecuária dos Cerrados. **Relatório técnico do projeto nipo-brasileiro de cooperação em pesquisa agrícola nos cerrados 1987/1992.** Planaltina: Embrapa/CPAC-JICA, pp.258-270, 1994.

PEREIRA, F.T.; MARQUES, M.G.; CARVALHO, D.D.C. Produção *in vitro* de escleródios de *Sclerotinia sclerotiorum* sob diferentes regimes de luz. **Revista Biociências,** v.22, n.1, p.56-60, 2016.

SMITH, M.E.; HENKEL, T.W.; ROLLINS, J.A. How many fungi make sclerotia?. **Fungal Ecology,** v.13, p. 211-220, 2015. https://doi.org/10.1016/j.funeco.2014.08.010

STEADMAN, J.R.; BOLAND, G. White mold. In: SCHWARTZ, H.F.; STEADMAN, J.R.; HALL, R.; FORSTER, R.L. **Compendium of bean diseases**. Saint Paul: American Phytopathological Society. pp.44-46, 2005.

VENTUROSO, L.R.; BACCHI, L.M.A.; GAVASSONI, W.L.; CONUS, L.A.; PONTIM, B.C.A. Relação de massa e localização do escleródio no solo com germinação carpogênica de *Sclerotinia sclerotiorum*. **Summa Phytopathologica**, v.40, n.1, p.29-33, 2014. https://doi.org/10.1590/S0100-54052014000100004

WENDLAND, A.; MOREIRA, A.S.; BIANCHINI, A.; GIAMPAN, J.S.; LOBO JUNIOR, M. Doenças do Feijoeiro. In: AMORIM, L.; REZENDE. J.A.M.; BERGAMIN FILHO, A.; CAMARGO, L.E.A. **Manual de Fitopatologia: Doenças das plantas cultivadas**. vol.2, 5.Ed. Ouro Fino: Agronômica Ceres, pp.383-396, 2016.

WILLETTS, H.J. Survival of fungal sclerotia under adverse environmental conditions. Biological Reviews of the Cambridge. **Philosophical Society**, v.46, n.3, p.387-407, 1971.

9. Powdery mildew - *Erysiphe polygoni*

Kingdom: Fungi
Phylum: Ascomycota
Order: Erysiphales
Family: Erysiphaceae

Etiology

This order comprises those fungi known as powdery mildew. One of the most common diseases in various crops, from vegetables, annual crops, fruit plants and even forest species. Individuals of the order Erysiphales are biotrophic parasites, that is, they need a living host to multiply. The parasitism mechanism is characterized by superficial colonization of the plant tissue. Initially, a germ tube is emitted from the conidium, which lengthens to produce at its end a specialized structure known as appressorium (VIELBA-FERNÁNDEZ et al., 2020). The appressorium allows the fungus to penetrate the cuticle and cell wall of plant leaf epidermis cells. Interestingly, the fungus leaves the plasma membrane of the host cell intact (EICHMANN & HÜCKELHOVEN, 2008).

As it is an order of the Phylum Ascomycota, the double nomenclature is present. To give an example, we can mention the genera *Erysiphe*, *Uncinula*, *Microsphaera*, *Podosphaera* and *Sphaerotheca*, all teleomorphs and which produce, in their anamorphic phases, asexual conidia of the genus *Oidium*.

The teleomorphic phase of fungi of this order is characterized by the production of spherical sexual ascoma known as cleistothecium, found on the host surface generally during autumn and winter. Morphological differentiation of teleomorphic genera is possible based on the number of asci per cleistothecium and the appendix formed. The genera *Sphaerotheca* and *Podosphaera* produce one asci per cleistothecium.

In the anamorphic phase, conidia of the order Erysiphales are produced in chains, in a basipetal form, that is, the youngest conidium is positioned at the base of the conidiophore. Thus, the conidia produced correspond to the anamorph *Oidium*, which are unicellular, hyaline, ovoid to cylindrical in shape and produced from short and unbranched conidiophores (MASSOLA JUNIOR, 2018). Regarding the dimensions of the anamorph, when found in bean plants, DENG et al. (2022) reported the following measurements: cylindrical and erect conidiophores measuring 44.2-76.0 x 8.0-10.3 µm and elliptical to ovoid conidia measuring 25.4-35.8 x 13.5-19.8 µm.

According to CAMPA & FERREIRA (2017), there is controversy regarding the causal agent of common bean powdery mildew, which has often been attributed to the fungus *Erysiphe polygoni* (SCHWARTZ et al., 2005), while some studies suggest that the causal agent would be more closed to *Erysiphe diffusa*. In other words, in phylogenetic studies, when analyzing ITS sequences of nuclear ribosomal DNA (nrDNA), ALMEIDA et al., (2008) found that *Erysiphe* sp. obtained from bean plants in Brazil (isolate EB2004, Genbank accession AY739109) showed a closer genetic relationship with *Erysiphe diffusa*. The *Erysiphe* fungus corresponds to the teleomorphic phase of the Phylum Ascomycota (Figure 16).

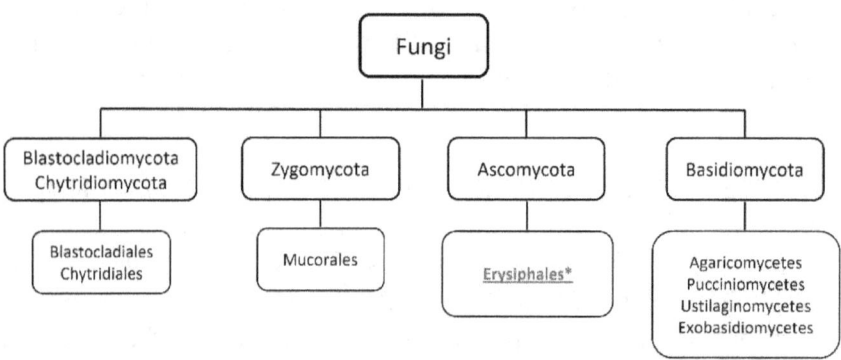

Figure 16. Schematic diagram shows the positioning of the fungus *Erysiphe polygoni* within the order Erysiphales, of the Fungi kingdom.

Symptomatology

The initial symptoms appear as small white spots similar to talc on the adaxial leaves surface (MURUBE et al., 2017), as can be seen in figure 17. The peak of symptoms is characterized by the stronger presence of a white powdery mass on the leaves adaxial face, which expand and fuse to form a layer that covers the entire leaf surface (TRABANCO et al., 2012; DENG et al., 2022). Infected leaves gradually curl downward, change color from pale yellow to brown, and ultimately leaf abscission occurs. Furthermore, the presence of white powdery mildew mycelium on the leaf surface causes a reduction in the photosynthetic rate (XAVIER et al., 2015).

Figure 17. Powdery mildew: leaves of *Vigna unguiculata* exhibiting the initial symptoms of powdery mildew, characterized by the appearance of small white spots similar to talc on the adaxial leaves surface.

Epidemiology
Climate conditions: The geographic distribution of this disease is increasing rapidly in different parts of the world, affecting large areas. This expansion has been related to changes in the global

climate, since populations of this pathogen evolve rapidly, presenting a high rate of variation due to the coexistence of sexual and asexual stages of the fungus, as well as the high dispersal capacity (CAMPA & FERREIRA, 2017). These characteristics in particular make powdery mildew a useful model for studying the effects of climate change on plant diseases (GLAWE, 2008). The damage caused by this disease is significant, especially if the fungus occurs before flowering under temperatures of 20 to 24°C, high humidity and a shaded environment (SCHWARTZ et al., 2005).

Dissemination: As it is a disease transmitted through the air, identification is a critical step to effectively prevent its spread, as well as minimize significant losses in yield and quality of produced seeds and grains (BINAGWA et al., 2021).

Survival: Attention should be paid to some species of the genus *Vigna*, more specifically *V. unguiculata* and *V. radiata*, as they are hosts of powdery mildew (DENG et al., 2022).

Control
Evasion: Several strategies are used to control powdery mildew, including adjusting of the planting date aiming to synchronize the crop with periods of maximum sun exposure (BINAGWA et al., 2021).

Immunization: The development of resistant common bean varieties is considered one of the most economical, efficient and ecological measures for managing powdery mildew (TRABANCO et al., 2012; MURUBE et al., 2017). When it comes to identifying the wild bean genome as a source of resistance to powdery mildew, BINAGWA et al. (2021) suggest that resistance to the causal agent of the bean powdery mildew involves a network of many constitutively co-expressed genes.

Therapy: Another measure consists in the application of fungicides (BINAGWA et al., 2021).

References

ALMEIDA, A.M.R.; BINNECK, E.; PIUGA, F.F.; MARIN, S.R.R.; RIBEIRO DO VALLE, P.R.Z.; SILVEIRA, C.A. Characterization of powdery mildews strains from soybean, bean, sunflower and weeds in Brazil using rDNA-ITS sequences. **Tropical Plant Pathology,** v.33, n.1, p.20-26, 2008. https://doi.org/10.1590/S1982-56762008000100004

BINAGWA, P.H.; TRAORE, S.M.; EGNIN, M.; BERNARD, G.C.; RITTE, I.; MORTLEY, D.; KAMFWA, K.; HE, G.; BONSI, C. Genome-Wide identification of powdery mildew resistance in common bean (*Phaseolus vulgaris* L.). **Frontiers in Genetics,** v.12, 673069, 2021. https://doi.org/10.3389/fgene.2021.673069

CAMPA, A.; FERREIRA, J.J. Gene coding for an elongation factor is involved in resistance against powdery mildew in common bean. **Theoretical and Applied Genetics,** v.130, p.849-860, 2017. https://doi.org/10.1007/s00122-017-2864-x

DENG, D.; SUN, S.; WU, W.; DUAN, C.; WANG, Z.; ZHANG, S.; ZHU, Z. Identification of causal agent inciting powdery mildew on common bean and screening of resistance cultivars. **Plants,** v.11, 874, 2022. https://doi.org/10.3390/plants11070874

EICHMANN, R.; HÜCKELHOVEN, R. Accommodation of powdery mildew fungi in intact plant cells. **Journal of Plant Physiology,** v.165, p.5-18, 2008. https://doi.org/10.1016/j.jplph.2007.05.004

GLAWE, D.A. The powdery mildews. A review of the world's most familiar (yet poorly known) plant pathogens. **Annual Review of Phytopathology,** v.46, p.27-51, 2008. https://doi.org/10.1146/annurev.phyto.46.081407.104740

MASSOLA JÚNIOR, N.S. Fungos fitopatogênicos. In: AMORIM, L.; REZENDE, J.A.M.; BERGAMIN FILHO, A. **Manual de Fitopatologia: princípios e conceitos**. vol.1, 5.Ed. Ouro Fino:

Agronômica Ceres, pp.107-142, 2018.

MURUBE, E.; CAMPA, A.; FERREIRA, J.J. Identification of new resistance sources to powdery mildew, and the genetic characterisation of resistance in three common bean genotypes. **Crop and Pasture Science,** v.68, p.1006-1012, 2017. https://doi.org/10.1071/CP16460

SCHWARTZ, H.; STEDMAN, J.; HALL, R.; FORSTER, R. **Compendium of bean diseases.** 2nd Edition. American Phytopathological Society. 2005. 109p.

TRABANCO, N.; PÉREZ-VEJA, E.; CAMPA, A.; RUBIALES, D.; FERREIRA, J.J. Genetic resistance to powdery mildew in common bean. **Euphytica,** v.186, p.875-882, 2012. https://doi.org/10.1007/s10681-012-0663-7

VIELBA-FERNÁNDEZ, A.; POLONIO, A.; RUIZ-JIMÉNEZ, L.; VICENTE, A.; PÉREZ-GARCÍA, A.; FERNÁNDEZ-ORTUÑO, D. Fungicide Resistance in Powdery Mildew Fungi. **Microorganisms**, v.8, 1431, 2020. https://doi.org/10.3390/microorganisms8091431

XAVIER, S.A.; MELLO, F.E.; CANTERI, M.G.; GODOY, C.V. Fotossíntese de folhas de soja infectadas por *Corynespora cassiicola* e *Erysiphe diffusa.* **Summa Phytopathologica,** v.41, n.2, p.156-159, 2015. https://doi.org/10.1590/0100-5405/1923

10. Fusarium wilt - *Fusarium oxysporum* f. sp. *phaseoli*

Kingdom: Fungi
Phylum: Ascomycota
Order: Hypocreales
Family: Nectriaceae

Etiology

Hypocreales is one of the most important orders of the Phylum Ascomycota, as it comprises fungi from the Nectriaceae and Clavicipitaceae families. Perithecia are produced on or immersed in a stroma. The more known teleomorphs are *Haematonectria*, *Nectria, Giberella* and *Calonectria*. Among these, special attention is given to the genus *Giberella*, which has species of the genus *Fusarium* as anamorph, which cause vascular wilt (CARVALHO et al., 2015) and are transmitted as seeds pathogens (CARVALHO et al., 2011), attacking seedlings (TOLÊDO-SOUZA et al., 2009) and as producers of mycotoxins (DING et al., 2024).

Regarding morphocultural characters, the colonies have a white border and a violet or purple center, with macroconidia of short to medium length, falciform to nearly straight, fine walls and 3 septa, and microconidia without septa, oval or reniform in shape, formed in false heads in short monophialides (PAULINO et al., 2022).

There is a great variation in the morphology of *F. oxysporum* colonies when grown on Potato-dextrose-agar (PDA) medium, making them very difficult to distinguish from *F. solani* and *F. subglutinans* (LESLIE & SUMMEREL, 2006). It turns out that *F. oxysporum* is a complex of species. In studies conducted by HA et al. (2023), a total of seven *F. oxysporum* isolates were obtained from red beans, mung beans, and Adzuki beans. The isolates presented microconidia that were slightly curved (3-4 septa) and had an oval to clavate shape. Sporodochia were generally absent, and when present, they were orange in color. Such characters corroborate with the descriptions reported by LESLIE & SUMMEREL (2006). It is essential that in addition to morphological characterization, molecular identification be done, as was the case in the work conducted by HA et al. (2023). In this work, the dimensions obtained were provided in supplementary material. Briefly, microconidia from three sporodochium-forming isolates (NC20-772, NC20-773 and NC20-779) showed results of a similar pattern, measuring 5.8-12.8 x 2.2-4.1 µm, while macroconidia were 31.4-51.5 x 3.1-6.2 µm.

The teleomorphic phase for the fungus *Fusarium oxysporum* (Figure 18) corresponds to the genus *Giberella*, which rarely occurs on the field associated with lesions and infected tissues. It is also important to comment about the term formae specialis, which refers to the fungal physiology in terms of its pathogenicity, that is, specialization in a specific host, in this case, the common bean plant, belonging to the genus *Phaseolus* sp.

As the fungus *F. oxysporum* has a high degree of host specificity, the evaluation of its grouping in *formae specialis* is carried out using pathogenicity tests. As these tests are not conclusive, molecular techniques have been used to help the researchers, as well as analyze the genetic diversity of populations (CRUZ et al., 2018). As an example, we can mention sequencing using translation elongation factor 1-alpha (TEF-1α), which is very accurate and informative at the species level for identifying *Fusarium* (O'DONNELL et al., 2015). In a study conducted by PAULINO et al. (2022), the *Fusarium oxysporum* f. sp. *phaseoli* complex was characterized by several tools, employing sequences from TEF-1α, beta-tubulin (TUB2), RNA polymerase II and calmodulin (CAL). The results showed that representative isolates of *F. oxysporum* f. sp. *phaseoli* were more aggressive and comprise a monophyletic group that separate from the previously reported *F. oxysporum* complex, which included other species of lesser severity such as *F. nirenbergiae* and *F. fabacearum*.

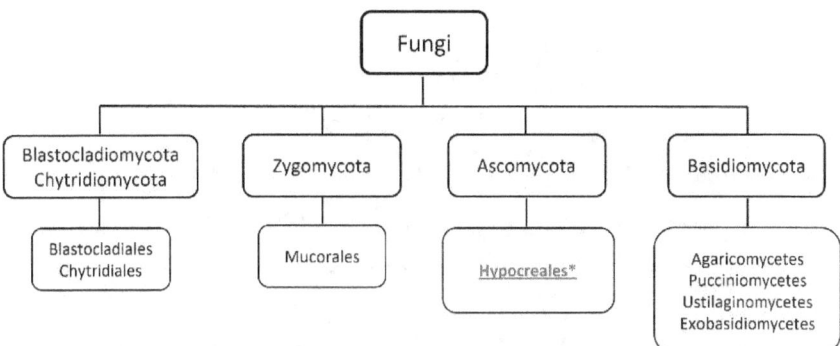

Figure 18. Schematic diagram shows the positioning of the fungus *Fusarium oxysporum* f. sp. *phaseoli*, within the order Hypocreales, from the Fungi kingdom.

Symptomatology

After entry of the pathogen through the root system, infections cause yellowing and wilting in plants (Figure 19), especially in the flowering and pod filling stages, wilting is irreversible and plants may eventually die prematurely (CARVALHO et al., 2015). Symptoms on the leaves are basically characterized by discoloration of the vascular tissue that leads to leaf necrosis and premature defoliation (FARIAS NETO et al. 2006). Among other symptoms, loss of turgidity is reported, starting from the lower leaves, which can manifest on only one side of the host plant (SCHWARTZ et al. 2005). When colonized, the vascular tissue acquires a reddish-brown color and can be easily seen after a bevel cut.

Figure 19. Common bean fusarium wilt: bean plants exhibiting yellowing and wilting symptoms.

Epidemiology

Climate conditions: Fusarium wilt may initially appear in isolated small outbreak and, after several seasons, spread throughout the area (ABAWI & PASTOR-CORRALES, 1990). The disease is favored by mild temperatures and high soil humidity (CARVALHO et al., 2015). Soil moisture is important for its survival, as the fungus survives as a saprophyte in the soil or in plant remains in the form of a resistant chlamydospores, which survive for a long time in the soil and in the host plant absence (CRUZ et al., 2018). Once the causal agent resides in the soil, penetration occurs via the root system and, once the plant is colonized, the pathogen can reach the seeds.

Dissemination: Seeds are important vehicles for phytopathogenic agents, which can cause a reduction in both germination and seedling vigor (CARVALHO et al., 2011).

Survival: The pathogen structures present on the seeds remain viable during the storage period and constitute the primary inoculum for the development of epidemics (SILVA et al., 2008). Therefore, infection can occur even at the seedling stage, damaging development and causing the appearance of abnormal seedlings (CARVALHO et al., 2011). The fungus *F. oxuysporum* can also survive in the soil in the form of infective chlamydospores.

Control

Exclusion: Attention must be paid to the seeds, and the exclusion of contaminated seeds is recommended.

Eradication: In the field of biological control, research has shown promising results, both in the treatment of seeds with *Trichoderma* strains (CARVALHO et al., 2014), and in soil application (CARVALHO et al., 2015). The successful management of fusarium wilt, leading to reduced yield loss in commercial fields, using competitive *Trichoderma* isolates has been reported in some research (SHALI et al., 2010).

Immunization: The disease is difficult to control, but there are some effective management strategies (HALL & NASSER, 1996). The use of resistant cultivars has proven to be an efficient method for controlling fusarium wilt (CARNEIRO et al. 2010). As an example, we can mention the black bean cultivar IAC Netuno, reported as resistant to fusarium wilt. This cultivar has a high grain yield potential of 2968.20 kg ha^{-1}, being an erect plant with an indeterminate Type II growth habit, 90-day cycle and recommended for crops in the state of São Paulo, Brazil (CHIORATO et al., 2020).

Therapy: Chemical fungicides are ineffective and are not recommended for vascular wilts (except for seed treatment), as these substances do not prevent root infection and phloem colonization by the pathogen (CARVALHO et al., 2015). Therefore, attention must be paid to seeds, because the use of healthy and treated seeds is one of the recommendations to prevent the transmission of diseases via seeds, in addition to contributing to a greater density of plants in the crop (CORRÊA et al., 2008).

References

ABAWI, G.S.; PASTOR-CORRALES, M.A. **Root rots of beans in Latin America and Africa: Diagnosis, research methodologies, and management strategies.** Centro Internacional de Agricultura Tropical (CIAT), Cali, 1990. 114p.

CARNEIRO, F.F.; RAMALHO, M.A.P.; PEREIRA, M.J.Z. *Fusarium oxysporum* f. sp. *phaseoli* and *Meloidogyne incognita* interaction in common bean. **Crop Breeding and Applied Biotechnology,** v.10, n.3, p.271-274, 2010. https://doi.org/10.1590/S1984-70332010000300014

CARVALHO, D.D.C.; LOBO JUNIOR, M.; MARTINS, I.; INGLIS, P.W.; MELLO, S.C.M. Biological control of *Fusarium oxysporum* f. sp. *phaseoli* by *Trichoderma harzianum* and its use for common bean seed treatment. **Tropical Plant Pathology,** v.39, n.5, p.384-391, 2014. https://doi.org/10.1590/S1982-6762014000500005

CARVALHO, D.D.C.; MELLO, S.C.M.; LOBO JÚNIOR. M.; SILVA, M.C. Controle de *Fusarium oxysporum* f.sp. *phaseoli in vitro* e em sementes, e promoção do crescimento inicial do feijoeiro comum por *Trichoderma harzianum*. **Tropical Plant Pathology,** v.36, n.1, p.028-034, 2011. https://doi.org/10.1590/S1982-56762011000100004

CARVALHO, D.D.C.; MELLO, S.C.M.; MARTINS, I.; LOBO JUNIOR. M. Biological control of Fusarium wilt on common beans by in-furrow application of *Trichoderma harzianum*. **Tropical Plant Pathology,** v.40, p.375-381, 2015. https://doi.org/10.1007/s40858-015-0057-1

CHIORATO, A.F.; CARBONELL, S.A.M.; BEZERRA, L.M.P.; SILVA, D.A.; GONÇALVES, J.G.R.; BENCHIMOL-REIS, L.L.; CARVALHO, C.R.L.; ESTEVES, J.A.F.; SANTOS, N.C.B.;

BARROS, V.N.P. IAC Netuno: A new black bean cultivar resistant to anthracnose and Fusarium wilt. **Crop Breeding and Applied Biotechnology,** v.20, n.3, e20442033, 2020. https://doi.org/10.1590/1984-70332020v20n3c37

CORRÊA, B.O.; MOURA, A.B.; DENARDIN, N.D.; SOARES, V.N.; SCHÄFER, J.T.; LUDWIG, J. Influência da microbiolização de sementes de feijão sobre a transmissão de *Colletotrichum lindemuthianum* Sacc. & Magn. **Revista Brasileira de Sementes,** v.30, p.156-163, 2008. https://doi.org/10.1590/S0101-31222008000200019

CRUZ, A.F.; SILVA, L.F.; SOUSA, T.V.; NICOLI, A.; PAULA JUNIOR, T.J.; CAIXETA, E.T.; ZAMBOLIM, L. Molecular diversity in *Fusarium oxysporum* isolates from common bean fields in Brazil. **European Journal of Plant Pathology,** v.152, p.343-354, 2018. https://doi.org/10.1007/s10658-018-1479-7

DING, Y.; MA, N.; HASEEB, H.A.; DAI, Z.; ZHANG, J.; GUO. W. Genome-wide transcriptome analysis of toxigenic *Fusarium verticillioides* in response to variation of temperature and water activity on maize kernels. **International Journal of Food Microbiology**, v.410, 110494, 2024. https://doi.org/10.1016/j.ijfoodmicro.2023.110494

FARIAS NETO, A.L.; HARTMAN, G.L.; PEDERSEN, W.L.; LI, S.; BOLLERO, G.A.; DIERS, B.W. Irrigation and inoculation treatments that increase the severity of soybean sudden death syndrome in the field. **Crop Science,** v.46, n.6, p.2547-2554, 2006. https://doi.org/10.2135/cropsci2006.02.0129

HA, M.S.; RYU, H.; JU, H.J.; CHOI. H. Diversity and pathogenic characteristics of the *Fusarium* species isolated from minor legumes in Korea. **Scientific Reports**, v.13, 22516, 2023. https://doi.org/10.1038/s41598-023-49736-4

HALL, R.; NASSER, L.C.B. Practice and precept in cultural

management of bean diseases. **Canadian Journal of Plant Disease,** v.18, p.176-185, 1996. https://doi.org/10.1080/07060669609500643

O'DONNELL, K.; WARD, T.J.; ROBERT, V.A.R.G.; CROUS, P.W.; GEISER, D.M.; KANG, S. DNA sequence-based identification of Fusarium: Current status and future directions. **Phytoparasitica,** v.43, n.5, p.583-595, 2015. https://doi.org/10.1007/s12600-015-0484-z

PAULINO, J.F.C.; ALMEIDA, C.P.; BARBOSA, C.C.F.; GONÇALVES, G.M.C.; BUENO, C.J.; HARAKAVA, R.; CARBONELL, S.A.M.; CHIORATO, A.F.; BENCHIMOL-REIS, L.L. Molecular and pathogenicity characterization of *Fusarium oxysporum* species complex associated with Fusarium wilt of common bean in Brazil. **Tropical Plant Pathology,** v.47, p.485-494, 2022. https://doi.org/10.1007/s40858-022-00502-3

SCHWARTZ, H.; STEDMAN, J.; HALL, R.; FORSTER, R. **Compendium of bean diseases.** 2nd Edition. American Phytopathological Society. 2005. 109p.

SHALI, A.; GHASEMI, S.; AHMADIAN, G.; RANJBAR, G.; DEHESTANI, A.; KHALESI, N.; MOTALLEBI, E.; VAHED, M. *Bacillus pumilus* SG2 chitinases induced and regulated by chitin, show inhibitory activity Against *Fusarium graminearum* and *Bipolaris sorokiniana*. **Phytoparasitica**, v.38, p.141-147, 2010. https://doi.org/10.1007/s12600-009-0078-8

SILVA, G.C.; GOMES, D.P.; KRONKA, A.Z.; MORAES, M.H. Qualidade fisiológica e sanitária de sementes de feijoeiro (*Phaseolus vulgaris* L.) provenientes do estado de Goiás. **Semina Ciências Agrárias,** v.29, p.29-34, 2008.

TOLÊDO-SOUZA, E.D.; LOBO JÚNIOR, M.; SILVEIRA, P.M.; CAFÉ FILHO, A.C. Interações entre *Fusarium solani* f. sp. *phaseoli* e *Rhizoctonia solani* na severidade da podridão radicular do feijoeiro. **Pesquisa Agropecuária Tropical,** v.39, n.1, p.13-17, 2009.

11. Root-knot nematode - *Meloidogyne* spp.

Kingdom: Animalia
Phylum: Nematoda
Class: Chromadorea
Superfamily: Tylenchoidea
Family: Meloidogynidae

Etiology

The systematics of nematodes has successive modifications over time, mainly due to the use of molecular techniques for the recognition of important species of plant nematodes. From 1998 onwards, Mark Blaxter's research group proposed a broad review of the nematode systematics, mainly encouraged by the phylogenetic advances obtained at the time. Thus, small proposals were made by some authors, until in 2006 this movement resulted in a currently accepted model, which was proposed by DECRAEMER & HUNT (2006).

When it comes to the morphology of *Meloidogyne*, specific patterns of the perineal region of females were used to identify species of the genus. However, this method presented a certain degree of subjectivity and lack of precision (FERRAZ & BROWN, 2016). Additionally, biochemical and molecular methods are recommended as auxiliary tools for identifying *Meloidogyne* species. As an example, we can mention the work of SILVA et al. (2021), who, when analyzing *M. enterolobii* occurring in beans, found that the analyzes of the perineal configuration were not conclusive. Identification was possible through electrophoresis of isoenzymes, that is, esterase and malate dehydrogenase phenotypes.

The female *Meloidogyne* differs from the male, since female has a pear shape and her dimensions are 0.4-1.3 mm long by 0.27-0.75 mm wide (AGRIOS, 2006).

Within the systematics proposed by DECRAEMER & HUNT (2006), the superfamily Tylenchoidea was the one that comprises a wide range of economically important nematodes, such as *Pratylenchus* and *Heterodera*, which belong to the families Pratylenchidae and Heteroderidade, respectively, and *Meloidogyne*, belonging to the family Meloidogynidae (Figure 20).

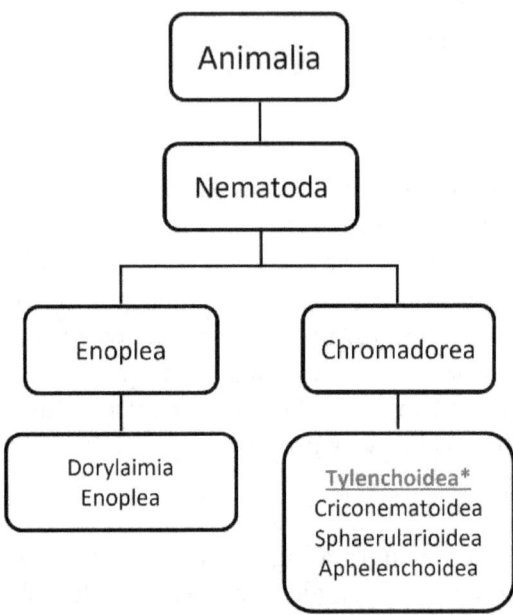

Figure 20. Schematic diagram shows the positioning of the genus *Meloidogyne* sp. within the systematics proposed by DECRAEMER & HUNT (2006), highlighting the superfamilies within the class Chormadorea, among them, the superfamily Tylenchoidea, to which it belongs.

Symptomatology

In the case of root-knot nematodes (Figure 21), attention is paid to the root system damage by the formation of giant cells, that is, the host cells become hypertrophied and become multinucleated (histological symptoms) (FERRAZ & BROWN, 2016). Such root damage, caused by the nematode, affects water and nutrient absorption. As consequence, reflex symptoms include reduced growth, small and yellow leaves, reducing the plant growth and plant production (SANTOS et al., 2012).

Figure 21. Common bean root-knot nematode: direct symptom (galls) characterized by thickening, of variable diameter, of the roots.

Epidemiology

According to JONES et al. (2013), the genus *Meloidogyne* comprises 98 plant parasitic species. The *Meloidogyne* life cycle includes six developmental stages: egg, four juvenile stages (J1-J4) and adult, with the only infective stage being J2 (GINÉ et al., 2021). Fundamentally, females of this genus lay eggs in gelatinous masses. Then, the juveniles of the first stage (J1) develop inside the egg. It is interesting to note that the first ecdysis occurs inside the egg, from which they emerge as juveniles of the second stage (J2). The three subsequent ecdyses occur when advancing between stages, as shown in figure 22.

At the J2 stage, juveniles are already able to penetrate the root system of the host plants and feed, causing the formation of galls on the roots. This occurs because *Meloidogyne* parasitism leads to the formation of "giant" cells (change in format due to hypertrophy and hyperplasia), which will provide sustenance to the juvenile, which is a female of the species. The female becomes sedentary, undergoes morphological changes and acquires, as a consequence of her process as a parasite, distension of her pear-shaped body (CHITWOOD & PERRY, 2009). The most common, frequent and harmful species of gall namtoid in common beans are *Meloidogyne incognita* and *Meloidogyne javanica* (FREIRE & FERRAZ 1977).

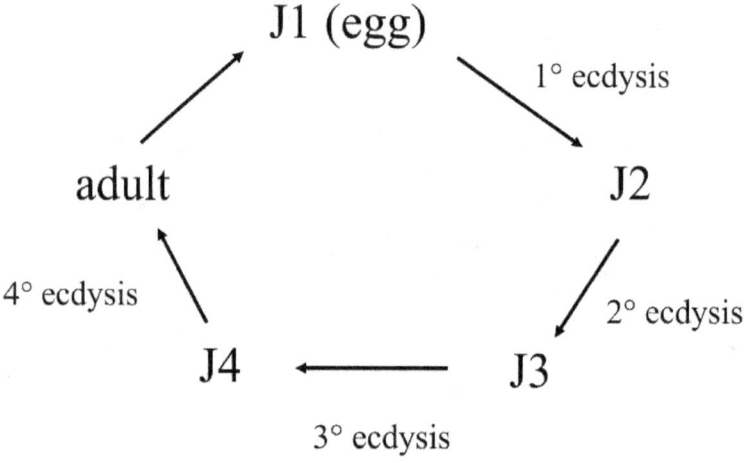

Figure 22. General scheme of the life cycle of the genus *Meloidogyne* sp., showing the existence of an ecdysis between each of the stages.

Climate conditions: The duration of the *Meloidogyne* life cycle is influenced by soil temperature, which is the main abiotic factor that determines the rate of nematode development. In *Phaseolus vulgaris* cv. Enana Nassau, GUINEA et al. (2021), found a smaller number of days (29 days) for *Meloidogyne incognita* to complete the life cycle from inoculation to the emergence of J2 when the average soil temperature was around 27.5°C. At lower temperatures, the number of days increased. It is worth mentioning that soil humidity and aeration are also other very influential abiotic factors in the nematode life cycle.

Dissemination: Another aspect that influences epidemiology is the location and dissemination of nematodes, which occur in greater abundance in the upper soil 15 to 30 cm. However, distribution across a cultivated area is generally irregular. The total distance covered by a nematode probably does not exceed a few meters per growing season (AGRIOS, 2005). This fact highlights the importance of equipment related to soil disturbance and dragging in the dissemination of nematodes to greater distances within planting areas. In addition to agricultural equipment, irrigation and drainage also spread nematodes on planting plots.

Survival: The mains survival strategy of *Meloidogyne* occurs by infecting the various crops it affects, such as soybeans and common beans. Optimal temperature conditions (23 to 30°C) for hatching, development and reproduction, allow the nematode to complete four to five generations within a cultivation cycle of these crops (90 to 120 days) (FERRAZ & BROWN, 2016).

Control
Nematode control is considered a difficult practice in agriculture, mainly because applications of chemical nematicides are often costly and aggressive to the environment (NUNES et al., 2010).

Evasion: Contaminated areas with high population density should be avoided.

Eradication: One management practice consists of crop rotation, because this practice can inhibit nematode reproduction. However, rotation is considered inefficient or limited by the scarcity of resistant materials with good agronomic characteristics (FERREIRA et al., 2012; FERNANDES et al., 2013).

Immunization: When it comes to managing plant diseases, the use of resistant and/or tolerant cultivars is considered a less costly and more efficient measure when cultivated in contaminated areas. The resistance of *Phaseolus vulgaris* to the two main species of *Meloidogyne* indicates the existence of accessions as possible sources of resistance in breeding programs, as was reported for *M.*

javanica (SANTOS et al., 2017) and *M. incognita* (PEDROSA et al., 2000). Finding sources of resistance has been a challenge. As an example, we can mention the study conducted by COSTA et al. (2019), in which evaluating 26 genotypes, all were susceptible to *M. incognita*, and only four were moderately resistant, when evaluating the *M. javanica* reproduction index.

References

AGRIOS, G.N. **Plant Pathology.** 5th ed. San Diego: Academic Press, 2005, 922p.

CHITWOOD, D.; PERRY, R. N. Reproduction, physiology and biochemistry. In: PERRY, R. N.; MOENS, M.; STARR, J. **Root-knot nematodes.** Wallingford: CABI, pp.182-200, 2009.

COSTA, J.P.G.; SOARES, P. L.M.; VIDAL, R.L.; NASCIMENTO, D.D.; FERREIRA, R.J. Reaction of common bean genotypes to the reproduction of *Meloidogyne javanica* and *Meloidogyne incognita*. **Pesquisa Agropecuária Tropical,** v.49, e54008, 2019. https://doi.org/10.1590/1983-40632019v4954008

DECRAEMER, W.; HUNT, D.J. Structure and classification. In: PERRY, R.N.; MOENS, M. **Plant Nematology,** Wallingford: CABI, pp.3-32, 2006.

FERNANDES, R.H.; LOPES, E.A.; VIEIRA, B.S.; BONTEMPO, A.F. Control of *Meloidogyne javanica* on common beans with *Bacillus* spp. isolates. **Revista Trópica: Ciências Agrárias e Biológicas,** v.7, p.76-81, 2013.

FERRAZ, L.C.C.B.; BROWN, D.J.F. **Nematologia de plantas: fundamentos e importância.** Manaus: Norma Editora, 2016, 251p.

FERREIRA, S.; GOMES, L.A.A.; MALUF, W.R.; FURTINI, I.V.; CAMPOS, V.P. Genetic control of resistance to *Meloidogyne incognita* race 1 in the Brazilian common bean (*Phaseolus vulgaris* L.) cv. Aporé. **Euphytica,** v.186, n.3, p.867-873, 2012.

https://doi.org/10.1007/s10681-012-0655-7

FREIRE, F.C.O.; FERRAZ, S. Nematóides associados ao feijoeiro, na Zona da Mata, Minas Gerais, e efeitos do parasitismo de *Meloidogyne incognita* e *M. javanica* sobre o cultivar "Rico 23". **Revista Ceres**, v.24, p.141-149, 1977.

GINÉ, A.; MONFORT, P.; SORRIBAS, F.J. Creation and validation of a temperature-based phenology model for *Meloidogyne incognita* on common bean. **Plants**, v.10:240, 2021. https://doi.org/10.3390/plants10020240

JONES, J.T.; HAEGEMAN, A.; DANCHIN, E.G.J.; GAUR, H.S.; HELDER, J.; JONES, M.G.K.; KIKUCHI, T.; MANZANILLA-LÓPEZ, R.; PALOMARES-RIUS, J.R.; WESEMAEL, W.M.L.; PERRY, R.N. Top 10 plant-parasitic nematodes in molecular plant pathology. **Molecular Plant Pathology,** v.14, n.9, p.946-961, 2013. https://doi.org/10.1111/mpp.12057

NUNES, H.T.; MONTEIRO, A.C.; POMELA, A.W.V. Use of microbial and chemical agents to control *Meloidogyne incognita* in soybean. **Acta Scientiarum Agronomy**, v.32, p.403-409, 2010. https://doi.org/10.4025/actasciagron.v32i3.2166

PEDROSA, E.M.R.; MOURA, R.M.; SILVA, E.G. Respostas de genótipos de *Phaseolus vulgaris* à meloidoginoses e alguns mecanismos envolvidos na reação. **Fitopatologia Brasileira**, v.25, p.190-196, 2000.

SANTOS, L.N.S.; ALVES, F.R.; BELAN, L.L.; CABRAL, P.D.S.; MATTA, F.P.; JESUS JUNIOR, W.C.; MORAES, W.B. Damage quantification and reaction of bean genotypes (*Phaseolus vulgaris* L.) to *Meloidogyne incognita* race 3 and *M. javanica*. **Summa Phytopathologica,** v.38, n.1, p.24-29, 2012. https://doi.org/10.1590/S0100-54052012000100004

SANTOS, L.N.S.; CABRAL, P.D.S.; NEVES, G.A.R.; ALVES,

F.R.; TEIXEIRA, M.B.; CUNHA, F.N.; SILVA, N.F. Multivariate statistics applied to the reaction of common bean plants to parasitism by *Meloidogyne javanica*. **Genetics and Molecular Research**, v.16, n.1, gmr16019420, 2017. https://doi.org/10.4238/gmr16019420

SILVA, R.V.; OLIVEIRA, J.O.; ÁVILA JÚNIOR, J.H.; LIMA, B.V.; MOREIRA, N.F. Occurrence of *Meloidogyne enterolobii* in common bean in southern Goiás State, Brazil
Ciência Rural, v.51, n.10, e20200403, 2021. https://doi.org/10.1590/0103-8478cr20200403

12. Cladosporium on seeds - *Cladosporium herbarum*

Kingdom: Fungi
Phylum: Ascomycota
Order: Capnodiales
Family: Davidiellaceae

Etiology

The fungus *Cladosporium herbarum* has tall, dark and erect conidiophores, irregularly branched at the apex and without nodules. The conidia are dark, have a prominent scar, without septa or with just one septum, ellipsoidal or lemon-shaped. Conidia have the following dimensions: 7.2-10.4 µm length x 3.6-4.2 µm width and average measures of 8.1 µm length x 4.0 µm width (GUIMARÃES & CARVALHO, 2014).

The fungus *Cladosporium herbarum* is an important seed pathogen of common bean crops, present in most of the untreated seed lots routinely analyzed. Like *Pseudocrecospora*, the *Cladosporium* fungus belongs to the Cpanodiales order (Figure 23). However, these two fungi do not share the same common family, since *Cladosporium* belongs to the Davidillaceae family and has the genus *Davidiella* as its teleomorph.

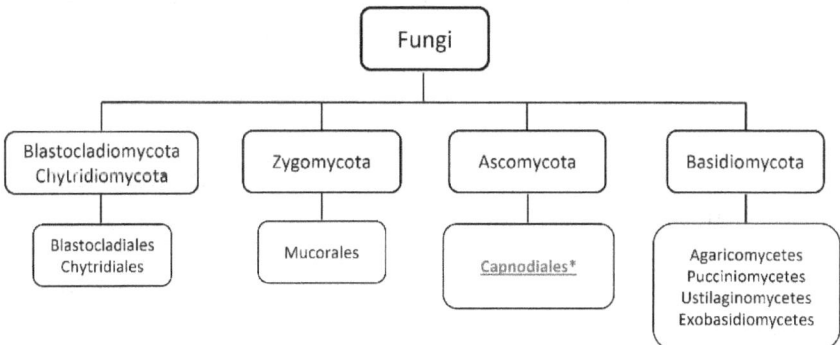

Figure 23. Schematic diagram shows the positioning of the fungus *Cladosporium herbarum* within the order Capnodiales, of the Fungi kingdom.

Symptomatology

Species of the genus *Cladosporium* occur on seeds, which will present spots on the tegument or greenish growths on the surface, mainly on the area corresponding to the embryo, more precisely the seed hilum, resulting in an undesirable appearance and consequent depreciation of seed lots (GUIMARÃES & CARVALHO, 2014).

Epidemiology

Climate conditions: When exposed to inadequate humidity and temperatures during storage, the growth of harmful fungi can occur (COSTA & SCUSSEL, 2022), resulting in loss of quality, viability and damage to the establishment of the initial crop stand (CARVALHO et al., 2011). It is also worth mentioning the collaboration of other factors for seed contamination, such as small openings on the surfaces caused by some species of insects and mechanical shocks resulting from harvesting, transport and storage (GUIMARÃES et al., 2018). All these factors must be observed aiming to minimizing the occurrence of *C. herbarum* in seeds.

Dissemination: When it comes to common bean cultivation, more than 50% of diseases are transmitted via seeds (MARINO & MESQUITA, 2009).

Survival: The fungus *Cladosporium* sp. survives associated with common bean seeds during storage, especially untreated seeds (GUIMARÃES et al., 2014).

Control

Eradication: To control this pathogen, seed treatment is recommended. Both chemical and biological treatment have shown good results. GUIMARÃES et al. (2014), evaluating *Trichoderma* isolates, obtained up to 77% suppression of *Cladosporium herbarum* in bean seeds cv. Perola, in addition to a high number of originated normal seedlings. In the same study, treatment with Carboxin + thiram reduced completely the incidence of *C. herbarum* on seeds. It is worth mentioning that replacing seed treatment with chemical products for treatment with biological products deserves to be considered due to the benefits to the environment that this practice can offer (CARVALHO et al., 2011). The incidence reduction of seed pathogens, through the employ of *Trichodema*, occurs because the antagonistic fungus competes with phytopathogenic fungi for the exudates released by the seeds during the germination process (HARMAN et al., 2004). Furthermore, the *Trichoderma* fungus has the ability for aggressively occupy the pathogen's establishment sites on the seeds and originated seedlings (GUIMARÃES et al., 2018).

References

CARVALHO, D.D.C.; MELLO, S.C.M.; LOBO JÚNIOR, M.; GERALDINE, A.M. Biocontrol of seed pathogens and growth promotion of common bean seedlings by *Trichoderma harzianum*. **Pesquisa Agropecuária Brasileira,** v.46, n.8, p.822-828, 2011. https://doi.org/10.1590/S0100-204X2011000800006

COSTA, L.L.F.; SCUSSEL, V.M. Toxigenic fungi in beans (*Phaseolus vulgaris* L.) classes black and color cultivated in the state of Santa Catarina, Brazil. **Brazilian Journal of Microbiology,** v.33, p.138-144, 2022. https://doi.org/10.1590/S1517-83822002000200008

GUIMARÃES, G.R.; CARVALHO, D.D.C. Incidência e caracterização morfológica de *Cladosporium herbarum* em feijão comum cv. 'Pérola'. **Revista Brasileira de Biociências,** v.12, n.3, p.137-140, 2014.

GUIMARÃES, G.R.; PEREIRA, F.S.; MATOS, F.S.; MELLO, S.C.M.; CARVALHO, D.D.C. Supression of seed borne *Cladosporium herbarum* on common bean seed by *Trichoderma harzianum* and promotion of seedling development. **Tropical Plant Pathology,** v.39, n.5, p.401-406, 2014. https://doi.org/10.1590/S1982-56762014000500007

GUIMARÃES, G.R.; PEREIRA, F.T.; MELLO, S.C.M. CARVALHO, D.D.C. *Trichoderma harzianum* no tratamento de sementes de *Cladosporium herbarum, Sclerotinia sclerotiorum* e no aumento de crescimento do feijoeiro no Brasil. **Caderno de Pesquisa,** v.30, n.02, p.28-37, 2018. https://doi.org/10.17058/cp.v30i2.6884

HARMAN, G.E.; HOWELL, C.R.; VITERBO, A.; CHET, I.; LORITO, M. *Trichoderma* species - opportunistic, avirulent plant symbionts. **Nature Reviews Microbiology,** v.2, p.43-56, 2004. https://doi.org/10.1038/nrmicro797

MARINO, R.H.; MESQUITA, J.B. Micoflora de sementes de feijão comum (*Phaseolus vulgaris* L.). provenientes do Estado de Sergipe. **Revista Brasileira de Ciências Agrárias,** v.4, p.252-256, 2009.